Charles Seale-Hayne Library
University of Plymouth
(01752) 588 588
LibraryandITenquiries@plymouth.ac.uk

Stormwater Modeling

STORMWATER MODELING

Donald E. Overton
DEPARTMENT OF CIVIL ENGINEERING
UNIVERSITY OF TENNESSEE
KNOXVILLE, TENNESSEE

Michael E. Meadows
ENVIRONMENTAL PLANNING ENGINEERS
KNOXVILLE, TENNESSEE

ACADEMIC PRESS   New York   San Francisco   London   1976

A Subsidiary of Harcourt Brace Jovanovich, Publishers

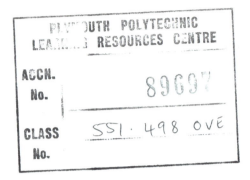
ACADEMIC PRESS, INC.
111 Fifth Avenue, New York, New York 10003

United Kingdom Edition published by
ACADEMIC PRESS, INC. (LONDON) LTD.
24/28 Oval Road, London NW1

Library of Congress Cataloging in Publication Data

Overton, Donald E.   Date
    Stormwater modeling.

    Includes bibliographical references and indexes.
    1.   Runoff–Mathematical models.   2.   Rain and rain-
fall–Mathematical models.   3.   Water–Pollution–
Mathematical models.   I.   Meadows, Michael E., joint
author.   II.   Title.
GB980.093   551.4'98   75-44762
ISBN 0–12–531550–3

PRINTED IN THE UNITED STATES OF AMERICA

Contents

Preface   xi

PART I   INTRODUCTION AND MODELING CONCEPTS

Chapter 1   Introduction and Modeling Concepts

1-1	Stormwater Defined	3
1-2	What Is a Mathematical Model?	4
1-3	The Systems Approach to Model Building	5
1-4	Systems Terminology and Definitions	6
1-5	The Modeling Approach	7
1-6	Example of Parametric Modeling	8
1-7	Example of Deterministic Modeling	9
1-8	Linkage between Parametric and Deterministic Modeling	10
1-9	Choice of Model Complexity	10
1-10	Stormwater Model Optimization by Objective Best Fitting	11
1-11	Sensitivity Analysis	13
1-12	Regionalization of Parameters	13
1-13	Range of Choice of Stormwater Models	14
	Problems	14
	References	15

PART II DETERMINISTIC MODELING

Chapter 2 Rainfall Excess

2-1   Introduction   19
2-2   Interception Models   19
2-3   Depression Storage Models   21
2-4   Soil Physics Models of Infiltration   22
2-5   Hydrologic Models of Infiltration   31
2-6   Evapotranspiration Models   37
      Problems   40
      References   40

Chapter 3 Overland and Open Channel Flow

3-1   Introduction   42
3.2   The Governing Equations of Motion   42
3-3   Kinematic and Dynamic Waves   45
3-4   Solution Techniques   47
      Problems   54
      References   56

Chapter 4 Kinematic Flow Approximation

4-1   Introduction   58
4-2   The Kinematic Approximation to Overland Flow   58
4-3   Kinematic Flow Number   61
4-4   Kinematic Flow on Long Impermeable Planes—
      the Rising Hydrograph   62
4-5   Time to Equilibrium   68
4-6   Equilibrium Depth Profile   68
4-7   The Falling Hydrograph   69
4-8   Model for a V-Shaped Watershed   70
4-9   Overland Flow on a Converging Surface   74
4-10 Overland Flow on a Cascade of Planes   76
4-11 Kinematic Shock   77
4-12 Kinematic Streamflow   79
4-13 Free Surface Storm Sewer Flow   81
      Problems   82
      References   85

Chapter 5 **Estimation of Time of Concentration Using Kinematic Wave Theory**

5-1 Introduction   87
5-2 Derivation of Time of Concentration   88
5-3 Time of Concentration for a Plane   88
5-4 Time of Concentration of a Cascade of Planes   89
5-5 Time of Concentration of a V-Shaped Watershed   90
5-6 Time of Concentration of a Converging Surface   91
5-7 A Concept of Lag Modulus   93
5-8 Estimation of Time of Concentration on Complex
     Catchments Where Stormwater Data Are Available   94
5-9 Summary   96
     Problems   96
     References   97

Chapter 6 **Examples of Deterministic Stormwater Modeling**

6-1 Introduction   98
6-2 A Model of Rural Stormwater   98
6-3 A Model of Urban Peak Runoff Design   108
6-4 Use of Stormwater Detention Basins for Peak
     Runoff Reduction   114
6-5 Hydrologic Impact of Storm Sewers   119
6-6 Stormwater Management Models   120
6-7 A Comparison of Urban Stormwater Models   125
     Problems   127
     References   127

Chapter 7 **Systems Approach to Deterministic Stormwater Modeling**

7-1 Introduction   129
7-2 Linear and Nonlinear Hydrologic Systems   129
7-3 Derivation of Response Function   131
7-4 Development of the Variable Response Model (VRM)   134
7-5 Examples of Application of VRM   146
7-6 Hydrologic Design of Stormwater Inlets   152
7-7 Conclusions   154
     Problems   154
     References   155

PART III PARAMETRIC MODELING

Chapter 8 History of Parametric Stormwater Modeling

8-1	Introduction	159
8-2	Parametric Modeling Prior to the High Speed Digital Computer	161
8-3	Parametric Modeling with High Speed Digital Computers	164
8-4	Present State of the Art	167
	Problems	167
	References	167

Chapter 9 Model Optimization Techniques

9-1	Need for Objective Best Fitting	169
9-2	Linear Least Squares	169
9-3	Nonlinear Least Squares	172
9-4	Principal Components Analysis	174
9-5	Stepwise Multiple Regression	186
9-6	Rosenbrock's Method	186
9-7	Pattern Search	187
9-8	Summary	189
	Problems	190
	References	191

Chapter 10 Evaluation of Effects of Urbanization and Logging on Stormwater

10-1	Introduction	192
10-2	TVA Stormwater Model	193
10-3	Model Tests	200
10-4	Simulated Effects of Forest Cutting on Stormwater	207
10-5	Simulated Effects of Urbanization on Stormwater	209
10-6	Model Limitations	213
	Problems	213
	References	213

Chapter 11 Evaluation of the Effects of Strip Mining on Streamflow

11-1	Introduction	215
11-2	TVA Daily Flow Model	216

11-3   Stanford Model   227
11-4   Conclusions   233
       Problems   236
       References   237

Chapter 12   **Sensitivity Analysis**

12-1   Need for Sensitivity Analysis   238
12-2   USGS Model   238
12-3   Model Optimization   241
12-4   A Case Study   241
12-5   Conclusions   243
       Problems   245
       References   245

Chapter 13   **Regionalization of Model Parameters**

13-1   Introduction   246
13-2   Regionalization of TVA Stormwater Model   246
13-3   Example Applications   250
13-4   Limitations of Regionalized Models   256
       Problems   256
       References   257

PART IV   STOCHASTIC STORMWATER MODELING

Chapter 14   **Stormwater Frequency Modeling**

14-1   Introduction   261
14-2   Return Period   262
14-3   Probability Density Functions   264
14-4   Plotting Positions   265
14-5   Best Fit Criteria   266
14-6   Outliers   272
14-7   Confidence Intervals—Reliability   272
14-8   Effects of Urbanization on Stormwater Frequency   274
14-9   Design Risk   277
14-10 Conclusions   279
       Problems   279
       References   280

PART V   STORMWATER QUALITY MODELING

Chapter 15   **State of the Art in Stormwater Quality**

15-1   Introduction   285
15-2   Stormwater as a Pollutant   286
15-3   Nonpoint Source   287
15-4   Effects of Pollutants on Receiving Streams and Lakes   288
15-5   Quality of Urban Runoff   291
15-6   Quality of Agricultural Runoff   294
15-7   Quality of Runoff from Mining Areas   298
15-8   Quality of Runoff from Forests and Woodlands   300
15-9   Conclusion   300
         Problems   301
         References   301

Chapter 16   **Simulating Pollutographs and Loadographs**

16-1   Introduction   303
16-2   Accumulation of Pollutants   304
16-3   Removal of Pollutants   311
16-4   Simulating Pollutographs and Loadographs   323
         Problems   342
         References   343

Chapter 17   **Development of Stormwater Quality Indices**

17-1   Introduction   345
17-2   "Consumers' Water Quality Index"   346
17-3   A Multivariate Approach   347
17-4   Principal Components Regression   349
17-5   Conclusions   350
         Problems   351
         References   351

Appendix I   353

Index   355

Preface

.

This book is an attempt to present the fundamentals of deterministic, parametric, and stochastic stormwater modeling. It is assumed that the reader or student will have a basic background in science or engineering; however, the authors are of the opinion that one can comfortably read and understand this treatise with a fundamental knowledge of calculus and differential equations. The book has been written with the intent of reaching an audience concerned primarily with evaluating the effects of land use on stormwater for the purpose of doing feasibility studies, planning, and/or design work. Hence, our attempt was to present only enough theory to provide the user with a basic understanding of the modeling concepts. If the reader desires to study further any of the concepts presented, references have been included which will provide a basis for a more rigorous development of the topics.

Land use activities are continuing with an increasing worldwide intensity. Assessment of the environmental impact of these activities is of great concern both before and after the fact. Urbanization, agricultural practices, coal strip mining, and logging operations are examples of land use activities that have allegedly contributed to flooding and stream water quality degradation. In order to develop defensible environmental impact statements associated with these activities, it is essential that the most scientifically based methodology be applied to the problems. Since little hydrologic data are available on smaller watersheds, it is becoming widely accepted that mathematical modeling is the only available means of making reliable predictions of the effects of land use changes on streamflow quantity and quality. The intent of the authors is to give a view of present methodology with illustrative examples.

The book is partitioned into two basic sections, one on quantity and the other on quality. However, if the focus of the reader is primarily on stormwater pollution, it must be carefully understood that the transport mechanism for water pollutants is the water itself. Therefore, one will not be able to generate stormwater pollutional loadings without first gaining a fundamental knowledge of the rainfall–runoff process.

Recent water quality studies have indicated that treatment of the waste water of a community will not be enough to achieve and maintain national, state, and local water quality standards. Efforts must also be made to utilize land resources in such a way as to minimize adverse effects on water quality. Problems of soil erosion and sedimentation, stormwater runoff, and changes in land-use patterns all have measurable effects on water quality. Any attempt to provide solutions to problems of water quality must begin with an adequate identification of the problems.

Solutions to water quality problems have generally been in the form of facilities—for transmitting and treating waste water. It has been determined that treatment, even advanced treatment, does not provide the total solution to water quality problems. Treatment systems must be supplemented by suitable land development practices and regulations. It follows that streamflow quality problems cannot be placed in a realistic perspective without delineating stormwater (nonpoint source) pollution relative to municipal and industrial (point source) pollution.

The contents of this book are an integration of the viewpoints that the authors have developed in the course of their academic and personal study, research, teaching, and consulting engineering experiences. The work, of course, represents our viewpoint of stormwater modeling and it is inevitable that we have missed or omitted material which others consider important. We would hope that readers will bring to our attention omitted work which they may consider important.

We would like to acknowledge the imaginative and creative work of the many scientists and engineers who have shaped the ideas in stormwater modeling. In particular, we would like to express our appreciation of the efforts and opinions of Colby V. Ardis, Jr., and Roger P. Betson, the Tennessee Valley Authority, Donald L. Brakensiek, US Agricultural Research Service, David A. Dawdy, US Geological Survey, and Willard M. Snyder and David A. Woolhiser, US Agricultural Research Service.

PART I | Introduction and Modeling Concepts

Chapter 1 | Introduction and Modeling Concepts

1-1 Stormwater Defined

Stormwater is the direct response to rainfall. It is the runoff which enters a ditch, stream, or storm sewer which does not have a significant base-flow component. The authors are not assuming that all stormwater reaches an open channel by the overland flow route, although conceptually many of the models do not have an interflow or base-flow component. In urban areas, this should be a realistic approach because of the high degree of imperviousness; however, in some rural watersheds an overland flow component may be nonexistent and direct storm response may be only near the stream and occur as seepage through the banks [1, 2].

As defined in this text, stormwater is associated with small upland or headwater watersheds where base flow is not a significant proportion of the total flow in the open channel during periods of rainfall. Hence, the attention in this text is directed principally at predicting watershed stormwater discharges as a function of land use and climate rather than predicting the water level along a river. The emphasis herein is upon the storm hydrograph rather than the stage hydrograph.

1-2 What Is a Mathematical Model?

A mathematical model is simply a quantitative expression of a process or phenomenon one is observing, analyzing, or predicting. Since no process can be completely observed, any mathematical expression of a process will involve some element of stochasticism, i.e., uncertainty. Hence, any mathematical model formulated to represent a process or phenomenon will be conceptual to some extent and the reliability of the model will be based upon the extent to which it can be or has been verified. Model verification is a function of the data available to test scientifically the model and the resources available (time, manpower, and money) to perform the scientific tests. Since time, manpower, and money always have finite limits, decisions must be made by the modelers as to the degree of complexity the model is to have, and the extensiveness of the verification tests that are to be performed.

The initial task of the modeler then is to make decisions as to which to use or to build, how to verify it, and how to determine its statistical reliability in application, e.g., feasibility, planning, design, or management. This decision-making process is initiated by clearly formulating the *objective* of the modeling endeavor and placing it in the context of the available resources on the project for fulfilling the objective.

If the initial model form does not achieve the intended objective, then it simply becomes a matter of revising the model and repeating the experimental verifications until the project objective is met. Hence, mathematical modeling is by its nature *heuristic* and *iterative*. The choice of model revisions as well as the initial model structure will also be heavily affected by the range of choice of modeling concepts available to the modeler, and by the skill which the modeler has or can develop in applying them.

Figure 1-1 is a schematic representation of the modeling process. The

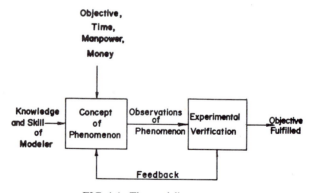

FIG. 1-1. The modeling process.

modeling process is not new but is nothing more than a modern expression of the classical scientific thought processes involved in the design of an experiment. What is very new and which was not available to Darwin or Euler is that today a very large number of concepts can be evaluated efficiently in a very small amount of time at a relatively small expense. The mechanisms which permit these evaluations are the high speed digital computer and a body of analytical techniques called *systems analysis*. To be effective, the modeler must therefore be knowledgable not only of the phenomenon itself but must also be knowledgable and skilled in computer science and in the discipline of systems analysis. Calculus and differential equations are the basic requirements for developing as a systems analyst.

1-3 The Systems Approach to Model Building

Dooge [3] has developed a good working definition of a system as being any structure, device, scheme, or procedure that interrelates an input to an output in a given time reference. The key concepts of a system are

1. A system consists of parts connected together in accordance with some sort of plan, i.e., it is an ordered arrangement.
2. A system has a time frame.
3. A system has a cause–effect relation.
4. A system has the main function to interrelate an input and output, e.g., storm rainfall and storm runoff.

Hence, the essence of systems analysis as applied to stormwater modeling is to interrelate rainfall (input) to stormwater (output) with a reliable model in a computationally efficient manner. The concept of the system is represented by the mathematical model which is an analytical abstraction of the real world and is by necessity an approximate representation of reality.

There are three basic steps which can be taken in model building:

1. Diagnosis
   a. Decide upon the objective of the modeling.
   b. Quantify the constraints placed upon the analysis.
      (1) time
      (2) money
      (3) available data
2. Formulate the problem mathematically
   a. Determine what decisions are to be made using the model.
   b. Identify critical components of model.

    c. Determine, through research of the open literature, how the components of the phenomenon have been viewed.

3. Construct the model

    a. Determine

        (1) data inputs, hydrologic and physical

        (2) required outputs

        (3) time domain of input and output

    b. Mathematically combine the components of the phenomenon in the context of governing physical laws.

1-4   Systems Terminology and Definitions

There has been an evolution of systems-modeling jargon, and it is important to review its main parts before proceeding to modeling work.

A *variable* has no fixed value (e.g., daily rainfall) whereas a *parameter* is a constant whose value varies with the circumstances of its application (e.g., Manning *n*-value).

The distinction between *linear* and *nonlinear* systems is of paramount importance in understanding the mechanism of mathematical modeling. A linear system is defined mathematically by a linear differential equation, the principle of superposition applies and system response is only a function of the system itself. An example of a linear system representation is the unit hydrograph model. A nonlinear system is represented by a nonlinear differential equation and system response depends upon the system itself and the input intensity. An example of a nonlinear system representation is the equation of gradually varied open channel flow. It is well known that real world systems are very nonlinear, but linear representations have often been made (e.g., the Streeter–Phelps stream dissolved oxygen model) because of lack of knowledge of the system or because of the pressures exerted by the resource constraint.

The *state* of a system is defined as the values of the variables of the system at an instant of time. Hence, if we know exactly where all of the stormwater is and its flowrate in a basin, then we know the state of the system. The state of the stormwater system is either determined from historical data or by assumption.

System *memory* is the length of time in the past over which the input affects the present state. If stormwater from a basin today is affected by the stormwater flow yesterday, the system (basin) is said to have a finite memory. If it is not affected at all, the system has no memory; and, if it is affected by storm flows since the beginning of the world, the system is said to have an infinite memory. Memory of surface water flow systems is mostly a function of antecedent moisture conditions.

A *time-invariant* system is one in which the input–output relation is not dependent upon the time at which the input is applied to the system. Stormwater flow systems are both *time-variant* and time-invariant depending upon size (acreage) and land use. The sediment load can also induce time-variance since channel roughness is directly affected.

A *lumped* variable or parameter system is one in which the variations in space either do not exist or have been ignored. The input is said to be lumped if rainfall into a system model is considered to be spatially uniform. Lumped systems are represented by ordinary differential equations and distributed systems are represented by partial differential equations.

A system is said to be *stochastic* if for a given input there is an element of chance or probability associated with obtaining a certain output. A *deterministic* system has no element of chance in it, hence for a given input a completely predictable output results for given initial and/or boundary values. A *purely random process* is a system with no deterministic component, and output is completely given to chance. A parametric or conceptual model does have an element of chance built into it since there will always be errors in verifying it on real data. It does therefore have a stochastic component. A "black box" model relates input to output by an arbitrary function, and therefore has no inherent physical significance.

1-5 The Modeling Approach

Stormwater models are needed in land use planning if the consequences of development strategies on the water resource are to be evaluated. Even where actual data collected under land use conditions similar to that being proposed are available, differences in site characteristics will tend to invalidate results that are simply transferred. However, in the typical situation, very little if any data are available for directly assessing the consequences of alternate development strategies. As a result, mathematical models must be employed in the planning process. These models are needed to account for differences in site characteristics and to simulate the consequences of alternate development schemes.

There are two conceptual approaches that have been used in developing stormwater models. An approach often employed in urban planning has been termed *deterministic* modeling or system simulation. These models have a theoretical structure based upon physical laws and measures of initial and boundary conditions and input. When conditions are adequately described, the output from such a model should be known with a high degree of certainty. In reality, however, because of the complexity of the stormwater flow process, the number of physical measures required would make a complete model intractable. Simplifications and approximations must therefore be made. Since there are always a number of unknown model coefficients or parameters that cannot be directly or easily measured, it is required that the model be verified. This means that the results from usable deterministic models must be verified by being checked against real watershed data wherever such a model is to be applied.

The second conceptual stormwater approach has been termed *parametric* modeling. In this case, the models are somewhat less rigorously developed and generally simpler in approach. Model parameters are not necessarily defined as measurable physical entities although they are generally rational. Parameters for these models are determined by fitting the model to hydrologic data usually with an optimization technique.

The two modeling approaches thus appear to be similar and indeed for some subcomponent models, the differences are relatively minor. The real difference between the two approaches lies in the number of coefficients or parameters typically involved. The typical deterministic model has more processes included and thus more coefficients to be determined. Because of the inherent interactions among processes in nature, these coefficients become very difficult to determine in an optimum sense. Because of interactions within the model, a range of values for various coefficients may all yield similar results. Hence, without rigorous model verification, the. output from a deterministic model are suspect. The parameters in a parametric model on the other hand, are determined by optimization (objective best fit criteria).

Both modeling approaches require data before the model can be employed. The significant difference lies where the data must be located. For the deterministic model, the data should be available at the site of the application. For the parametric model, this latter requirement can be avoided by employing a two-step approach. The model can be fitted to data at locations where it is available in order to obtain optimum model parameters. These parameter values can then be correlated with the physical characteristics of the catchment or watershed. When this is done over a geographic area, the model is said to be *regionalized*. Once regional relationships between the site characteristics and model parameters are developed, it then becomes possible to measure the site characteristics at locations where water resource data are unavailable and to predict reliably the model parameters and hence, make scientifically based predictions.

There are advantages to both modeling approaches. If observed rainfall and storm hydrographs are available, then both approaches can be employed in the development of the stormwater system process models.

1-6 Example of Parametric Modeling

Because of a wealth of hydrologic data in the TVA region, parametric models have been developed by the Tennessee Valley Authority [4, 5] for certain critical components of the system. These TVA models have been utilized in land use studies in the TVA region. The TVA example is a clear demonstration of the need for hydrologic data if parametric models are to be regionalized.

1-7 Example of Deterministic Modeling

The typical approach used in developing hydrologic system planning models is the deterministic or simulation model. The EPA "Storm Water Management Model" [6] is but one example of a host of these types of models that have been developed. The problem with the complete simulation model lies in the data required to define many of the boundary conditions. Since many of these measures are not available at the application site (since they necessitate research-level data gathering procedures), the results obtained from research studies in other cities and even other countries are often used to assign boundary condition values to the area of study. The simulated results obtained from such models at an application site, even when a degree of verification is achieved, are still suspect because of bias incorporated by the modeler and because boundary values determined at other cities are not necessarily applicable.

The components of stormwater systems modeling that are most difficult to assign boundary values for in the simulation models are those that quantify nonpoint sources of streamflow quality and quantity in urban and rural subareas. For example, a most difficult problem is quantifying the change in streamflow quantity and quality resulting from building a subdivision on an agricultural area. If, as in the typical simulation approach, estimates of a water quality constituent are considered only after urbanization occurs and these estimates are based upon boundary values measured in other cities, then the results of the bias of the modelers and the inappropriate data are incorporated into the findings.

To be objective, the urban planning models must be capable of reliably estimating the streamflow quality and quantity as it was prior to urbanization, and as it will be under alternate development strategies. Only when this is done can the impact of urbanization be evaluated. For example, simply identifying that pollution is associated with urbanization is not sufficient since rural streams and even streams in pristine areas are also polluted using accepted quality criteria (at least during flood periods). Urban planning should aim only at minimizing the impact of development upon water quantity and quality. Any schemes to improve water quality or reduce flooding beyond that of a natural stream can hardly be economically or environmentally justified.

1-8   Linkage between Parametric and Deterministic Modeling

Parametric and deterministic stormwater models can be and should be complimentary. Parametric stormwater models such as the TVA model or the USGS model are lumped system representations, and they have the capability of assessing the gross effects of land use on runoff, but they do not have the capability of assessing the sensitivity of internal distributions of land use on runoff. Deterministic stormwater models such as the EPA [6] model is a distributed system representation and it can simulate the transport mechanism at the source of runoff production to the basin outlet. The EPA [6] model can also simulate stormwater flow through storm and combined sewers.

Hence, if the hydrologic and physical data are available for the study site, both modeling approaches can be effectively utilized. The parametric model can be utilized as the most scientific basis for predicting basin storm hydrographs on a regional basis, and the deterministic model can be utilized to investigate various land use scenarios on runoff and to simulate the transport mechanism including quality constituents. For the most reliable results, the two model types should be correlated. Further, as the deterministic approach provides information and improvements for the parametric approach, so will the parametric approach feedback information to indicate where further detailed specification is needed and where areas of the problem are most in need of further study.

1-9   Choice of Model Complexity

The modeler must choose how complex a mathematical system representation should be made. As pointed out above, this choice is principally dictated by the project objectives, the knowledge and skill of the modeler, and resource constraints.

If a highly complex mathematical representation of the system under study is made, either parametric or deterministic, then the risk of not representing the system will be minimized but the difficulty of obtaining a solution will be maximized. Much data will be required, programming effort and computer time will be large, and the general complexity of the mathematical handling may even render the problem formulation intractable. Further, the resource constraints of time, money, and manpower may be exceeded. Hence, the modeler must determine the proper degree of complexity of the mathematical model such that the best problem solution will result and the effort will meet the project constraints. Conversely, if a greatly simplified mathematical model is selected or developed, the risk

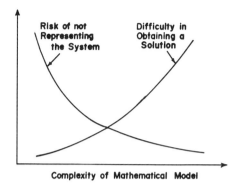

FIG. 1-2. The trade-off diagram.

of not representing the system will be maximized but the difficulty in obtaining a solution will be minimized. The main point here is that the modeler must make a decision from the range of choice of models available or from the models which could be built. But, as pointed out in Fig. 1-1, refinements in the model can be made by the modeler and indeed this is usually true.

Figure 1-2 is called the "trade-off diagram" because it illustrates the consequences of the decision of how complex the model should be. If after preliminary verification, it is determined that the initially chosen model is either too complex or not complex enough, then the modeler may move along the abscissa scale in Fig. 1-2 and experiment with another degree of complexity. This modeling effort should continue until the project objective is attained within the resource constraint.

1-10 Stormwater Model Optimization by Objective Best Fitting

Since parametric models are conceptual, a set of unknown coefficients or parameters will appear in the mathematical formulation. The parameter values in the model are experimentally determined in the verification procedure. Intuitively, the proper coefficient values would produce the best fit or linkage between storm rainfall (input) and the stormwater hydrograph (output). There is an instinctive temptation, which has appeared in modeling literature, to derive model parameters from observed storms by trial-and-error "eyeballing" best fit procedures. There are certain distinct and far reaching disadvantages associated with this approach to model verification. They are

1. If the model is of average complexity, about four or five parameters, then there are a very large number if not an infinite set of coefficients which will produce essentially the same fit. Hence, a large operational bias is induced into the modeling process and attributing physical significance to and regionalizing the model parameters may be precluded.

2. If the goodness of fit between the model and the observed stormwater hydrograph is not quantified, the "eyeballing" technique itself induces another operational bias and the same negative effects as (1) above will result.

3. The trial-and-error process is very time consuming and inefficient. Time constraints will permit but a relatively few number of computer trials.

This last point will be demonstrated using an example. Assume that there are five parameters in the stormwater model being utilized, and that a high, low, and middle range value of each of the five parameters is to be tested on a single storm. Further, it is assumed that it will take 10 sec/computer run to try each combination of parameters and that the computer costs are $300/hr. The number of computer runs will be 3^5 or 243. At 10 sec/run, the computer time would be 0.675 hr and it would cost $202.50 for each storm. It would then be necessary to plot all 243 hydrographs and pick the "eyeball" best fit. The modeler would be literally buried in output. Also, the choice of parameter values used in the tests would induce another operational bias since the modeler will select parameter values on the basis of previous experience and on a judgment which is considered reasonable.

By contrast, if an objective best fit technique, such as pattern search, is built into the computer model, optimum parameter values for each storm event analyzed will be derived by the exact same criteria thereby minimizing operational bias. Further, the computer output is minimized, the computer time per storm event analyzed would be approximately 10 sec, and the computer bill would then be $0.83 as opposed to the $202.50 in the "eyeball" experiment. More computer programming and debugging is involved in building in the parameter optimization scheme but this will quickly be offset by the high efficiency in production running.

For further contrast, trial-and-error fitting of 10 parameters with three assumed values for one storm event would require 164 hr of computer time and cost $49,000. Trial and error of 20 parameters with three assumed values would require 3.5×10^9 trials, 1100 yr of computer time and nearly $3 billion!

The conclusion here is that parameter optimization in parametric stormwater models is achieved scientifically and economically by utilization of objective best fit criteria rather than by a trial-and-error "eyeballing" process. However, there is a limit to the size and complexity of a parametric model which can be optimized using an objective best fit criteria. The US Geological Survey model [7] has nine parameters and has been successfully optimized. The Stanford [9] model, of which there are many versions, has at least 20 parameters and optimization by objective best fit criteria has not been done and is seemingly intractable.

1-11 Sensitivity Analysis

Model verification is not complete without a thorough sensitivity analysis. Once the calibrated parameters are arrived at by a best fit procedure, sensitivity analysis proceeds by holding all parameters constant but one, and perturbating the last one such that variation of the objective function (measure of fit between the observed storm hydrograph and the fitted model) can be examined. If small perturbations of the parameter produce large changes in the objective function, the system is said to be sensitive to that parameter. This gives a measure of how accurate that parameter must be estimated if the model is to be used in prediction. If the objective function is not sensitive to the perturbated parameter, then the parameter need not be accurately estimated in prediction. If the system is extremely insensitive to the perturbated parameter, the parameter and its associated system component may be redundant and could be deleted from the model.

1-12 Regionalization of Parameters

The effectiveness of parametric stormwater models will be measured, in the long run, by the confidence modelers will have in their ability to estimate model parameters on basins which have no hydrologic data for calibrating the model being utilized. A high level of confidence could be achieved if enough bench mark watersheds with hydrologic data were available for analysis. Optimized model parameters for each basin could then be related to physiographic, land use, and climatic characteristics of the study basin. This would permit an interpolation and extrapolation of the results to ungauged basins within the study region at some specified confidence level.

There have been very few reported attempts at parameter regionalization in the open literature. Primarily, this has been the result of a general lack of hydrologic data and research support money. The Stanford [9] model was originally regionalized in California and the TVA [5] stormwater model was regionalized in the Tennessee Valley. Neither of these regionalization experiments resulted in a very high level of confidence. Much work needs to be done, but data and money are needed.

1-13 Range of Choice of Stormwater Models

Linsley [10] reported a critical review of the models available for urban stormwater in 1971. He reviewed numerous models, real or apparent, and concluded that most had been developed for rural or big river systems and none were directly applicable to urban systems. Since that time, Papadakis and Pruel [11] compared the newly created University of Cincinnati urban stormwater model with the EPA [6] model and concluded that both models were very sensitive to infiltration estimates. The TVA [4, 5] models have both been reported since Linsley's [10] review.

The purpose of this book is not to perform a critical review of stormwater models. As pointed out in the preface, the purpose of this book is twofold. First, it is intended to develop the basic principles which have been or could be used in both deterministic and parametric stormwater modeling, and second it is intended to demonstrate the use of stormwater models with real world examples and problems. The following chapters are intended to fulfill these objectives.

Problems

1-1. What is system analysis?

1-2. What is a system constraint?

1-3. Discuss the difference between deterministic and parametric modeling.

    (a) When would each be appropriate?

    (b) How can they both be utilized in the same basin?

1-4. What logical steps are involved in developing and verifying a mathematical model of stormwater?

1-5. Do the following differential equations represent

    (a) linear or nonlinear systems?

    (b) lumped or distributed systems?

    (c) time-variant or time-invariant systems?

    (d) homogeneous or nonhomogeneous systems?

        (1) $(dQ/dt) + (Q/K) = f_1(t)$

        (2) $(V\partial V/\partial X) + (\partial V/\partial t) + (g\partial y/\partial X) = g(S_0 - S_f)$

        (3) $(dQ/dt) + [Q^{1/2}/K(t)] = f_2(t)$

        (4) $h_{xx} + h_{yy} + h_{zz} = f(t)$

        (5) $dL/dt = -KL$

(6) $\quad Y(dY/dt) + Y = t$

(7) $\quad (dY/dx) + (Y/x) = x^2$

1-6. What is the major tradeoff which must be made in model building?
1-7. Why is model optimization using an objective best fit criteria necessary?
1-8. How may confidence in use of parametric and deterministic models be established?
1-9. What are the basic data requirements for a parametric model? A deterministic model?

References

1. Betson, R. P., What is watershed runoff? *J. Geophys. Res.* **69,** No. 8, 1541–1552 (April 1964).
2. Ragan, R. M., An experimental investigation of partial area contributions, Intern. Assoc. Sci. Hydrol., General Assembly of Berne, pp. 241–249, 1968.
3. Dooge, J. C. I., Linear theory of hydrologic systems, US Dept. of Agr., Agr. Res. Service, Tech. Bull. No. 1468, 327 pp. (October 1973).
4. Tennessee Valley Authority, Upper Bear Creek experimental project—A continuous daily-streamflow model, Tennessee Valley Authority, Division of Water Control Planning, Knoxville, Tennessee, 99 pp. (February 1972).
5. Tennessee Valley Authority, Storm hydrographs using a double-triangle model, Tennessee Valley Authority, Division of Water Control Planning, Knoxville, Tennessee, 112 pp. (January 1973).
6. Chen, W. W., and Shubinski, R. D., Computer simulation of urban stormwater runoff, *J. Hydrol. Div.* **97,** HY2, 289–301 (February 1971).
7. Dawdy, D. R., and O'Donnell, T., Mathematical models of catchment behavior, *J. Hydrol. Div.* **91,** HY4, 123–139 (July 1965).
8. Green, R. F., Optimization by the pattern search method, Res. Paper No. 7, Tennessee Valley Authority, Division of Water Control Planning, Knoxville, Tennessee, 73 pp. (January 1970).
9. Linsley, R. K., and Crawford, N. H., Computation of a synthetic streamflow record on a digital computer, *Intern. Assoc. Sci., Hydrol. Pub.* **51,** 526–538, 1960.
10. Linsley, R. K., A critical review of currently available hydrologic models of analysis of urban stormwater runoff, Hydrocomp International (August 1971).
11. Papadakis, C. N., and Pruel, H. C. Methods for determination of urban runoff, *J. Hydrol. Div.* **99,** HY9, 1319–1335 (September 1973).

PART II | Deterministic Modeling

PART II | Deterministic Modeling

Chapter 2 | Rainfall Excess

2-1 Introduction

The first major component needed in building a stormwater model is rainfall excess which is sometimes called direct runoff. Abstractions or losses are subtracted from input rainfall resulting in the rainfall excess which must be routed to the basin outlet. Deterministic methods for the needed routing are presented in Chapter 3.

The losses to be abstracted from rainfall are shown in Fig. 2-1 and are:

1. Interception losses
2. Evapotranspiration losses
3. Depression or pocket storage
4. Infiltration losses

There are two basic approaches to modeling rainfall excess. First, each of the four losses shown above can be modeled and linked together or a single model could be developed in a lumped fashion. Examples of both approaches will be shown in this chapter.

2-2 Interception Models

Storm rainfall which is caught by vegetation prior to reaching the ground is referred to as interception losses. The amount intercepted is a

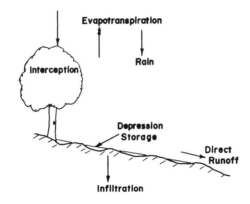

FIG. 2-1. Losses of rainfall which produce effective rainfall.

function of (a) the species, age, and density of the vegetation, (b) character of the storm, and (c) the season of the year. It has been estimated that 10–20% of the rainfall which falls during the growing season is intercepted and returned to the hydrologic cycle by evaporation [1].

Except for dense vegetation such as a forest, water losses by interception are not pronounced. Kittredge [2] reported that the average annual interception loss by Douglas fir stands in western Oregon and Washington was about 24%. Interception losses in urbanized areas should be minimal because of the sparcity of vegetation and because the higher rain intensities associated with storms would probably result in a substantial portion of any intercepted rain to ricochet to the ground. Interception losses develop during the early portion of the storm and the rate of interception rapidly approaches zero.

The only interception model of worthy note in the literature has been reported by Kittredge [2] and Chow [3] and is of the form

$$L = S + KEt \qquad (2\text{-}1)$$

where L is the volume of water intercepted in inches, S the interception storage retained against the forces of wind and gravity, K the ratio of surface area of intercepting leaves to horizontal projection of the area, E the amount of water evaporated per hour during the rain period, and t the time in hours. Because of the general lack of interception data, the common practice is to deduct the estimated volume entirely from the storm as part of the initial abstraction.

The reported values of S vary from 0.01 to 0.05 in. and a representative value of E is about 0.002 in./hr. Using 9 value of S of 0.01 in., the estimated amount of intercepted water for an urban area with 10% foliage in the first hour of a storm would be

$$L = 0.01 \text{ in.} + 0.10 \, (0.002 \text{ in./hr})(1 \text{ hr}) = 0.01002 \text{ in.} \qquad (2\text{-}2)$$

and every hour after this initial hour, there would be 0.0002 in. Hence, in urban areas or any areas barren of vegetation, interception is negligibly small.

2-3 Depression Storage Models

Water caught in the small voids or swales which is held until it infiltrates or evaporates is called depression storage. Because of the wide variability of the depressions and the general lack of experimental data, a generalized relation or model of the process does not exist.

Linsley et al. [4] reported that the volume of water stored by depressions, V, at any given instant of time after the beginning of rainfall could be approximated by

$$V = S_d[1 - \exp(-kP_e)] \qquad (2\text{-}3)$$

where S_d is the maximum storage capacity of the depressions, P_e the rainfall minus infiltration, interception, and evaporation, and k, the constant equivalent to $1/S_d$.

Hicks [5] has reported values of depression storage of 0.10 in. for clay, 0.15 in. for loamy soil, and 0.20 in. for sandy soils. Viessman [6] reported a range of depression storage of 0.06–0.11 in. for four small impervious drainage areas. He showed a strong relation between S_d and slope as shown in Fig. 2-2.

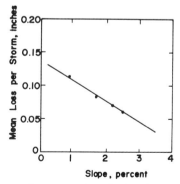

FIG. 2-2. Depression-storage loss versus slope for four impervious drainage areas. (After Viessman [6].)

Since there is a finite upper limit on S_d, it follows from Eq. (2-3) that the larger the rainstorm, the less significant depression storage will be in storm-water model calculations. Hence, if attention is directed at the more intense rainstorms in urban areas, then depression storage would not be an important component. By contrast, if attention is directed at less intense storms on land with a small amount of imperviousness, then depression storage would be a very important component model. This is an example of the general principle developed in Chapter 1, that the objective of the modeling effort is strongly interrelated to the model structure itself.

For a more complete treatment of depression storage the reader is referred to Viessman *et al.* [1].

2-4   Soil Physics Models of Infiltration

There are two basic deterministic approaches to infiltration, the soil physics and hydrologic approach. An excellent treatise on soil water physics has been written by Hillel [7] and a comparison of hydrologic infiltration models was reported by Overton [8].

The soil physics approach would seemingly constitute a more reliable model of infiltration and much research on the flow phenomenon has been done in the last 15 yr. However, the two main drawbacks to the use of this type of model are that there is a very large data requirement needed for the model, and the computations required are extensive. Further, infiltration is not purely a one-dimensional vertical flow process but depends upon surface slope and the ratio of vertical to horizontal soil flow properties (conductivity) [9]. Hence, the modeler must decide if the additional time, labor, and expense associated with such a highly deterministic infiltration model is worth the additional accuracy relative to that which could be obtained using a hydrologic infiltration model.

Derivation of Governing Equations

A one-dimensional model will be developed which will be made up of two parts, a conservation of mass equation and an equation of motion. Conservation of mass simply states that inflow minus outflow is equal to the change in storage in the soil element under study. A soil element is shown in Fig. 2-3. The volume rate of inflow entering the element is

$$\text{inflow} = v \, dx \tag{2-4}$$

where v is the specific discharge. The volume flowrate leaving the soil element is

$$\text{outflow} = [v + (\partial v/\partial z) \, dz] \, dx \tag{2-5}$$

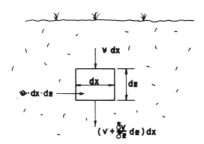

FIG. 2-3. Derivation of conservation of mass for an unsaturated soil mass.

The volume of moisture stored in the element at any given instant of time is

$$\text{volume} = \theta \, dx \, dz \qquad (2\text{-}6)$$

where θ is the ratio of moisture volume in the soil element relative to the total volume of the soil element. Then the conservation of mass equation becomes

$$v \, dx - [v + (\partial v/\partial z) \, dz] \, dx = \partial(\theta \, dx \, dz)/\partial t \qquad (2\text{-}7)$$

Equation (2-7) reduces to

$$(\partial v/\partial z) + (\partial \theta/\partial t) = 0 \qquad (2\text{-}8)$$

The equation of motion is based upon Darcy's laboratory experiments in saturated soil moisture flow [10]. The permeameter experiment is shown in Fig. 2-4. Darcy found that the flowrate v was proportional to the loss in hydraulic head H over the soil column relative to length of the soil column. This is expressed as

$$v \sim \frac{H_2 - H_1}{l} \qquad (2\text{-}9)$$

The hydraulic head in saturated soil moisture flow is equal to the sum of

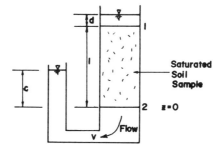

FIG. 2-4. Darcy's permeameter experiment in saturated soil–moisture flow.

the elevation or gravity head and the pressure head (which is assumed to be hydrostatic). Assigning $z = 0$ at point 2, the hydraulic head loss over the column is

$$H_2 - H_1 = c - (l + d) \tag{2-10}$$

Assigning a constant of proportionality called hydraulic conductivity K, Darcy's law for vertical saturated flow is

$$v = -K \, dH/dz \tag{2-11}$$

Darcian flow is viscous in nature and the minus sign in Eq. (2-11) denotes that flow is in the direction of decreasing head. The head loss is attributable to viscous or shear drag. Further, the velocity of the fluid is assumed to be changing very slowly both in space and in time, which explains the absence of acceleration terms in Eq. (2-11). Conductivity is a predictable function of both the fluid (density and viscosity) and the porous media (porosity, size gradation, packing, etc.).

The Darcy model has been extrapolated to unsaturated flow, but hydraulic head is a function of gravity head and pore pressure, tension, or suction. The cohesiveness of the porous media and the degree of saturation determine the tension. The forces acting under unsaturated flow are illustrated in Fig. 2-5.

Tension is much larger than gravity when the soil has a low water content. Hence, in the initial stages of infiltration, the flow process is controlled by tension. As the pores fill up, tension is reduced and gravity becomes important. When tension is exactly equal to gravity, a condition called field capacity is reached. This is the maximum moisture content because tension, which acts equally in all directions, is no longer large enough to hold additional water in the pores against the force of gravity. The only way for complete saturation to occur is for the groundwater table to back up into the unsaturated zone.

The soil moisture flow process during wetting under conditions where the soil–air interface is always saturated yields a typical infiltration curve shown in Fig. 2-6. Initially, tension is very large and accounts for the high infiltration rates. As field capacity is approached, tension is substantially

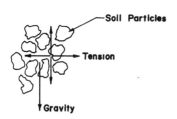

FIG. 2-5. Forces acting during unsaturated soil–moisture flow.

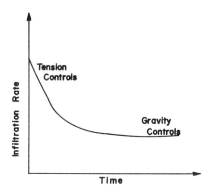

FIG. 2-6. Typical field infiltration curve where soil–air interface is saturated.

reduced and gravity begins to control the process. The leveling off of the infiltration curve is often explained as being a result of the field capacity effect. The system would be at an equilibrium state since the moisture content in the pores is not changing.

Hydraulic head is then equal to tension ψ plus gravity z.

$$H = \psi + z \tag{2-12}$$

and Darcy's law as applied to vertical unsaturated soil moisture flow is

$$v = -K\,\partial(\psi + z)/\partial z = -K[(\partial\psi/\partial z) + 1] \tag{2-13}$$

Conductivity exhibits a large variation with water content θ and in space. However, if conductivity is considered to be *constant* everywhere in the media, the media is said to be isotropic and homogeneous. If at a point in the media, conductivity is considered to vary in the x, y, z domain, the media is said to be anisotropic.

The equation of motion, known as Richard's equation, is formed by combining Eq. (2-13) with Eq. (2-8).

$$\partial\theta/\partial t = (\partial/\partial z)\{K[(\partial\psi/\partial z) + 1]\} \tag{2-14}$$

This governing equation is usually placed in the form of a diffusion type equation by making the transformation

$$\partial\psi/\partial z = (\partial\psi/\partial\theta)(\partial\theta/\partial z) \tag{2-15}$$

and defining diffusivity D as

$$D = K(\partial\psi/\partial\theta) \tag{2-16}$$

Equation (2-14) then becomes

$$\partial\theta/\partial t = (\partial/\partial z)[D(\partial\theta/\partial z) + K] \tag{2-17}$$

The solution of Eq. (2-17) for unsaturated soil moisture flow then depends upon specifying diffusivity D and conductivity for the field site.

The Moisture Characteristic

The amount of water remaining in the soil is a function of the tension or suction head ψ. This function is usually measured experimentally and is graphically represented by a curve known as the "soil–moisture characteristic," so named by Childs [11]. No satisfactory theory or set of field experiments exists which adequately describes or predicts the moisture characteristic for soils [7]. Several empirical relations have been proposed [12, 13] but they apply to a very limited range of soils.

The problem is further compounded by *hysteresis* in that the relation between suction and moisture is not unique and single valued. The relation depends upon whether the moisture flow process is undergoing wetting, i.e., *sorption*, or is undergoing draining, i.e., *desorption*. These relations are shown in Fig. 2-7. Generally, for a given water content, suction is lower during wetting than during drainage and little hysteretic loops can occur between the main hysteretic loops. This depends upon moisture availability. The hysteretic effect is attributable to (1) geometric nonuniformities of individual pores, (2) variations in contact angle in wetting and drainage, (3) entrapped air, and (4) swelling [7]. Conductivity likewise exhibits a hysteretic effect with moisture content typified by the relation shown in Fig. 2-8. From the modeling viewpoint, the natural flow process is very complicated and there is a general lack of field data to base a parameter prediction upon. Also, the soil moisture flow component system has a finite memory meaning that the solution of Richard's equation will normally depend heavily upon the initial state.

These complications have led investigators to seek solutions by simplifying Richard's equation. One such simplification is to assume that D and

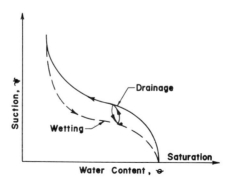

FIG. 2-7. The soil–moisture characteristic for wetting and drainage.

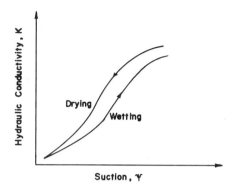

FIG. 2-8. Typical hydraulic conductivity–suction relation.

K are invariant with water content or in space. Equation (2-17) reduces to

$$\partial\theta/\partial t = D\,\partial^2\theta/\partial z^2 \tag{2-18}$$

The solution of Eq. (2-18) depends upon the boundary conditions and initial values. The known solutions have been catalogued by Eagleson [14] and hence will not be repeated here. The solutions are all in the form of exponentials and error functions. An examination of these solutions should lead the reader to the conclusion that even these greatly simplified solutions are quite complex for model handling. In spite of these complications, a watershed model has been developed and field tested by Smith and Woolhiser [15] which numerically integrates Richard's equation linked with the surface runoff equations. Their results will be presented in Chapter 6.

Green and Ampt Model

A conceptual model utilizing Darcy's law was proposed by Green and Ampt [16]. Referring to Fig. 2-9, the following assumptions were made in the derivation:

(1)   There exists a distinct and precisely defineable wetting front;

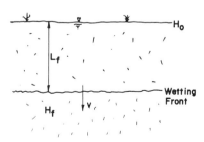

FIG. 2-9. Green and Ampt model.

(2)   Suction at this wetting front H_f remains essentially constant regardless of time and position;

(3)   Behind the front, the soil is uniformly wet and of constant conductivity K; hence,

(4)   Wetting front is thus viewed as a plane separating a uniformly wetted infiltrated zone from a totally uninfiltrated zone;

(5)   In effect, this supposes the K versus θ relation to be discontinuous, i.e., to change abruptly, at the value of suction prevailing at the wetting front.

By Darcy's law, infiltration rate becomes

$$v = K(H_0 - H_f + L_f)/L_f \tag{2-19}$$

Assuming the ponding depth H_0 negligible, we define

$$\Delta H_p = H_0 - H_f \quad \text{(constant)} \tag{2-20}$$

The cumulative infiltration F is

$$dF/dt = v = d(\Delta\theta L_f)/dt \tag{2-21}$$

where $\Delta\theta = \theta_t - \theta_i$, and θ_t is transmission zone wetness during infiltration, and θ_i is initial soil profile wetness which prevails beyond the wetting front. Combining Eqs. (2-19)–(2-21), we obtain

$$\Delta\theta \frac{dL_f}{dt} = K \frac{\Delta H_p + L_f}{L_f} \tag{2-22}$$

which can be integrated to form

$$\frac{Kt}{\Delta\theta} = L_f - \Delta H_p \ln\left(1 + \frac{L_f}{\Delta H_p}\right) \tag{2-23}$$

By using $F = \Delta\theta L_f$, Eq. (2-23) can be placed in terms of cumulative infiltration

$$t = \frac{\Delta\theta \, \Delta H_p}{K} \left\{ \frac{F}{\Delta\theta \, \Delta H_p} - \ln\left(1 + \frac{F}{\Delta\theta \, \Delta H_p}\right) \right\} \tag{2-24}$$

Little success has been reported in utilizing the Green and Ampt model primarily because the effective wetting front suction must be found by experiment. As a rule of thumb for infiltration into an initially dry soil, it may be about 0.1 atm [7].

EXAMPLE 2.1   For an initially dry soil with the following experimentally determined values: $\Delta H_p = 0.1$ atm, $\Delta\theta = 0.05$, and $K = 1$ ft/day, estimate how long it would take for 1 in. of rain water to infiltrate into this soil profile. One atmosphere of pressure corresponds to 33.9 ft of water.

Using Eq. (2-24), we obtain

$$t = \frac{(0.05)(0.1 \text{ atm})}{1 \text{ ft/day}} \left\{ \frac{1 \text{ in.}}{(0.05)(0.1 \text{ atm})} - \ln\left[1 + \frac{1 \text{ in.}}{(0.05)(0.1 \text{ atm})}\right] \right\}$$

$$= \frac{0.169 \text{ ft}}{1 \text{ ft/day}} \left\{ \frac{(1/12) \text{ ft}}{0.169 \text{ ft}} - \ln\left[1 + \frac{(1/12) \text{ ft}}{0.169 \text{ ft}}\right] \right\}$$

and

$$t = 0.0156 \text{ days} = 22.4 \text{ min}$$

The average infiltration rate is approximately 1 in./(22.4/60)hr = 2.68 in./hr.

A Variation of Green and Ampt Model

The Green and Ampt [16] model could be made more general by assuming that the ponding $H_0(t)$ is time variant by including rainfall intensity $i(t)$ as input to the system as shown in Fig. 2-10. Infiltration rate would be

$$v = K \frac{H_0(t) - (p/\gamma) + L_f}{L_f} \tag{2-25}$$

where p/γ is the pressure head at the wetting front. If the soil is at field capacity, then

$$|H_0(t) + L_f| \gg |p/\gamma| \tag{2-26}$$

and

$$v = K \frac{H_0(t) + L_f}{L_f} \tag{2-27}$$

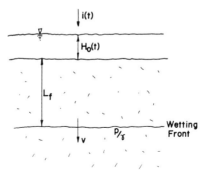

FIG. 2-10. A variation of Green and Ampt model.

Conservation of mass states that

$$i(t) - v = dH_0/dt \qquad (2\text{-}28)$$

and if rain rate is steady, Eq. (2-28) becomes

$$H_0 = it - F \qquad (2\text{-}29)$$

Equation (2-27) then becomes

$$\frac{dF}{dt} = K\left(\frac{it - F + L_f}{L_f}\right) \qquad (2\text{-}30)$$

Remember that $F = \Delta\theta \cdot L_f$; then Eq. (2-30) can be written

$$\frac{1}{2}\frac{dL_f^2}{dt} + \frac{K(\Delta\theta - 1)}{\Delta\theta} L_f = \frac{Kit}{\Delta\theta} \qquad (2\text{-}31)$$

Equation (2-31) would have to be integrated numerically; however, for the special case where rainfall has stopped, the right-hand side of Eq. (2-31) would be zero and the solution would be

$$L_f = \frac{K(1 - \Delta\theta)}{\Delta\theta} t \qquad (2\text{-}32)$$

where

$$\Delta\theta \cdot L_f \leq H_0 \qquad (2\text{-}33)$$

EXAMPLE 2.2   Four feet of water is ponded on a soil with conductivity equal to 1 ft/day and the water table is 8 ft beneath the surface.

If the porosity of the soil is 0.4, estimate (a) the time it will take for the wetting front to reach the water table, (b) the depth of ponding remaining (if any), and (c) the average infiltration rate. Assume that the difference between wetness in transmission zone and the wetness of the initial soil profile is one-half of the porosity.

(a)   Using Eq. (2-32), $\Delta\theta = 0.2$, hence

$$t = \frac{0.20\,(8\ \text{ft})}{(1\ \text{ft/day})(0.8)} = 2.0\ \text{day}$$

(b)   The amount of infiltrated water F is, by Eq. (2-33),

$$F = 0.2 \times 8\ \text{ft} = 1.6\ \text{ft}$$

hence, the amount of ponding remaining at 48 hr is

$$H_0(48) = (4-1.6)\,\text{ft} = 2.4\,\text{ft}$$

(c)   The average infiltration rate $\bar{v}$ is

$$\bar{v} = \frac{d(\Delta\theta \cdot L_f)}{dt} = K(1 - \Delta\theta) \tag{2-34}$$

and

$$\bar{v} = (1\,\text{ft/day})(0.8) = 3.5\,\text{ft/day}$$

The infiltration rate, as seen by Eq. (2-34) is implicitly assumed to be steady.

Other Infiltration Models

There are numerous other infiltration models which have either been derived from Richard's equation for certain simplifying assumptions or have been conceptually derived such as the Green and Ampt model. A complete literature review of these models can be found in Hillel [7]. Since the purpose of this book is to develop theory and example solutions of deterministic models, the literature review will not be repeated here.

2-5   Hydrologic Models of Infiltration

Because of the complexities in dealing with soil physics models, several hydrologic models have been reported in the literature. Two which have been utilized in stormwater modeling will be reported here. Conceptually, hydrologic models of infiltration are based upon a die-away of rate until a final rate is reached. It is often stated that at this point a confining layer has been reached.

Horton's Model

Horton [17] proposed an infiltration equation which he derived from work–energy principles. It is

$$f = f_c + (f_0 - f_c)\,e^{-kt} \tag{2-35}$$

where f is the infiltration rate at time t in inches per hour, f_0 and f_c are the initial and final infiltration rates in inches per hour, and k is the infiltration constant which is allegedly a function of soil and vegetation. The infiltration rates are "capacity" rates indicating that the soil–air interface must be saturated at all times. In practical terms this means that it is assumed that rainfall rate is always greater than infiltration capacity rates, and hence some ponding will always result. This is a major disadvantage in the use of Horton's model since rainfall rates are highly variable and therefore will fall below the values of capacity infiltration rates.

There has been little field experimentation with Horton's model which makes it very difficult to estimate the parameters f_0, f_c, and k for soils where there are no hydrologic data available. However, Musgrave [18] performed nationwide double-ringed infiltrometer experiments on hundreds of soils and correlated soil type with final rate of infiltration f_c. These results, which will be described below, were incorporated into the U.S. Soil Conservation Service Engineering Handbook [19]. Little experimental information exists on the parameters f_0 and k, but it is generally known that estimates of capacity infiltration rates are very sensitive to small errors in estimating these parameters, especially k.

Some field experiments of Watson [20] on silty clay loam to heavy clay with an initially dry grassy surface resulted in the following best fit values of the parameters: $f_0 = 1.75$ in./hr, $f_c = 0.5$ in./hr, and $k = 4.93$ hr^{-1}. These parameter values vary widely with soil type and initial moisture conditions, but the experiment of Watson [20] may serve to give the reader an indication of the range of values the parameters could have.

It is important to recognize that infiltration, as described by Horton's model, becomes essentially equal to the final infiltration rate in a small amount of time. Referring back to the experiment of Watson, the final rate was very nearly reached after 45 min.

EXAMPLE 2.3   In the field experiment of Watson [20], use Horton's model to estimate the volume of infiltration in inches which accumulated in the soil profile after 45 min.

Equation (2-35) must be integrated to arrive at the accumulated volume F.

$$F = f_c t + \frac{f_0 - f_c}{k}(1 - e^{-kt})$$

Then

$$F = 0.5 \ (\text{in./hr})(0.75 \ \text{hr}) + \frac{1.25 \ (\text{in./hr})}{4.93}\{1 - \exp[-4.93(0.75)]\}$$

$$= 0.622 \ \text{in.} \tag{2-36}$$

For this soil at the specified antecedent condition, the capacity infiltration rate became approximately constant after infiltrating 0.62 in. of rainwater in 45 min. Hence, we may conclude that for this soil condition, infiltration estimates are very important for high-intensity short-duration rainstorms;

but for long-duration storms, capacity infiltration rates can be considered to be essentially constant. Urban catchments and small rural watersheds are flashy, i.e., responsive to high-intensity short-duration rainstorms.

Holtan [21] proposed a conceptual model of infiltration backed by substantial field experimentation. He recognized from soil physics that as the pores fill up, infiltration rate dies away and approaches a final rate. The final rate of infiltration f_c was associated with the gravity force at field capacity. The model was then formulated to relate capacity infiltration rate to the volume remaining, F_p, as

$$f = aF_p^n + f_c \qquad (2\text{-}37)$$

The parameters a and n were determined experimentally from infiltrometer plot data. The exponent was found to be about 1.4 for all plots studied and the coefficient varied from 0.2 to 0.8 for the soil-cover complexes studied.

The problem with Eq. (2-37) is that an exponent of 1.4 does not permit it to be integrated in order to obtain a time distribution of capacity rates. Hence, a reexamination by Overton [8] showed that the integration was permitted using an exponent of 2 without a significant loss in accuracy from the original work of Holtan [21]. The integration proceeds by recognizing that the volume available to be infiltrated at the beginning of rain, $F_p(0)$, is equal to the available water capacity (AWC) minus the initial soil moisture (IM), greater than hydroscopic water.

$$F_p(0) = \text{AWC} - \text{IM} \qquad (2\text{-}38)$$

Then the relation between F_p and accumulated volume of infiltration F is

$$F_p(t) = F_p(0) - F(t) \qquad (2\text{-}39)$$

Equation (2-37) then becomes

$$dF/dt = a\{F_p(0) - F\}^2 + f_c \qquad (2\text{-}40)$$

which integrates to

$$F = F_p(0) - (f_c/a)^{1/2} \tan\{(af_c)^{1/2}(t_c - t)\} \qquad (2\text{-}41)$$

and the capacity rate equation is

$$f = f_c \sec^2\{(af_c)^{1/2}(t_c - t)\} \qquad (2\text{-}42)$$

where t_c is the time to constant rate of infiltration expressed as

$$t_c = (af_c)^{-1/2} \tan^{-1}\{(a/f_c)^{1/2} F_p(0)\} \qquad (2\text{-}43)$$

The parameter a was found to vary widely with antecedent moisture. An example of this variation is shown in Fig. 2-11 which are experimentally

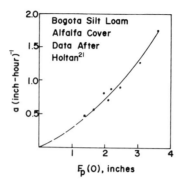

FIG. 2-11. Variation of parameter a with available volume of infiltration. (Data after Holtan [21].

determined values of a for a plot of Bogota silt loam with an alfalfa cover. Generally, a would be highest at the wilting point and least at field capacity. Thus, Holtan's parameter varies with initial conditions just as the Horton k-value does; the variation is substantial.

There are two main advantages of the Holtan model over the Horton model. First, there is a more reliable basis for estimating the parameters in Holtan's model. The results of a nationwide soil survey by the Agricultural Research Service [22] resulted in a catalog of AWC for hundreds of soils. Antecedent moisture would have to be estimated for any model; hence, only one parameter a need be estimated for Holtan's model, whereas two, f_0 and k, need be estimated before using Horton's model. Second, for rates less than capacity, the general form of the differential equation (2-40) may be integrated numerically, and with the aid of Eqs. (2-38) and (2-39), a complete moisture accounting scheme is readily available.

Overton [8] has also shown that there are distinct mathematical and hydrologic similarities amongst many of the infiltration equations.

Horton's model has been utilized recently in deterministic urban stormwater models, such as the University of Cincinnati model, whereas Holtan's model has been used in deterministic rural watershed models such as developed by Huggins and Monke [23].

EXAMPLE 2.4   Using the data given in Example 2.3, estimate the parameters a and t_c in the Holtan–Overton infiltration model and compare the Horton and Holtan–Overton models with the field data.

Solve for a in Eq. (2-40)

$$a = \frac{f - f_c}{\{F_p(0) - F\}^2}$$

at $t = 0$, $F(0) = 0$ and $f_0 = 1.75$, hence

$$a = \frac{1.75 - 0.50}{(0.622)^2} = 3.23 \text{ (in.-hr)}^{-1}$$

From Eq. (2-43),

$$t_c = (3.23 \times 0.50)^{-1/2} \tan^{-1}\{(3.23/0.50)^{1/2} \times 0.622\}$$

$$= 0.79 \text{ hr} = 47.5 \text{ min}$$

which agrees favorably with the field observation. The initial infiltration rate is

$$f_0 = 0.50 \times \sec^2\{[3.23(0.50)]^{1/2} \times 0.79\} = 1.73 \text{ in./hr}$$

Both models are plotted for comparison as shown in Fig. 2-12.

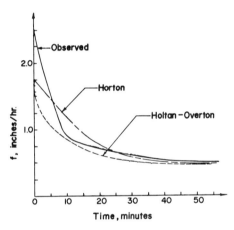

FIG. 2-12. Comparison of infiltration models with field experiment of Watson [20].

As an alternative to linking interception, depression storage, and in-filtration models, the U.S. Soil Conservation Service (SCS) [19] developed a direct runoff model which lumped all losses except evapotranspiration into a single initial abstraction. The model, which is referred to as the "curve number," correlates the rainfall-direct runoff relations as a function of soil type, land use, and hydrologic condition. The model was based upon a large amount of plot and small watershed runoff data as it existed in the early 1950s.

For simple storms (high intensity and of short duration), the retention relative to the potential maximum retention S bears the following relation:

$$\frac{P - Q - I_a}{S} = \frac{Q}{P - I_a} \tag{2-44}$$

where P is storm rainfall in inches, Q the direct storm runoff or effective rainfall in inches, and I_a the initial abstraction in inches. The concept behind this method appears to be that for a given basin soil and land use condition, there is a maximum possible retention and as storm rainfall increases, storm runoff will increase as defined by Eq. (2-44).

Storm runoff can be solved for from Eq. (2-44)

$$Q = \frac{(P - I_a)^2}{(P - I_a) + S} \tag{2-45}$$

An empirical relation for initial abstraction was inserted into Eq. (2-45)

$$I_a = 0.2S \tag{2-46}$$

and Eq. (2-45) becomes

$$Q = \frac{(P - 0.2S)^2}{P + 0.8S} \tag{2-47}$$

which is the relation used in the SCS method of estimating direct runoff from storm rainfall.

The runoff equation was placed in graphical form with the main parameter being watershed retention S, where S was related to curve number CN, as

$$S = \frac{1000}{CN} - 10 \tag{2-48}$$

Equations (2-47) and (2-48) are plotted in Fig. 2-13 which is taken from the SCS Engineering Handbook [19]. Figure 2-13 was verified to some extent using all available plot and small watershed data in the early 1950s.

The relation between CN and soils, land use and hydrologic condition for average antecedent moisture conditions is shown in Tables 2-1 and 2-2, also taken from the SCS Engineering Handbook [19]. Hydrologic soil type corresponds to Musgrave's [18] classification scheme. The SCS Handbook [19] has a list of hundreds of soils and their associated grouping (an A-soil is sandy and a D-soil is clay).

EXAMPLE 2.5   A rural 250 acre watershed has been classified as a B-soil by Musgrave [18]. In 1970 the watershed was entirely in pasture and in good hydrologic condition. The next year 100 acres were converted to small grain. The planted area was contoured. Estimate the change in runoff due to the land use change for a 3 in. storm rainfall.

From Table 2-1, the CN in 1970 was 61. After the land use change, the CN for the 150 acres of pasture was still 61, whereas the CN for the 100 acres in grain changed to 75. Then the weighted CN in 1971 was

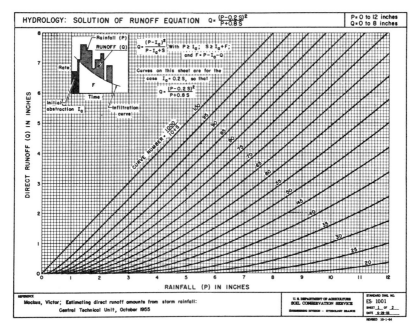

FIG. 2-13. SCS runoff curve numbers. (After US SCS [19].)

$$
\begin{array}{ll}
\text{pasture} & 150 \times 61 = 9150 \\
\text{grain} & 100 \times 75 = 7500
\end{array}
$$

$$\text{weighted } CN = 16{,}650/250 = 67$$

From Fig. 2-13, the predicted storm runoff volumes before and after the land use change are

$$
\begin{array}{ll}
\text{after} = & 0.58 \text{ in.} \\
\text{before} = & -0.36 \text{ in.}
\end{array}
$$

$$\text{increase} = -0.22 \text{ in. or } 61\%$$

Although the prediction technique of the SCS is entirely specified, little information concerning its reliability has been reported in the open literature.

2-6 Evapotranspiration Models

It is possible to analyze the evapotranspiration process and to formulate mathematical models for use as predictive tools. These models are highly complex and require a large amount of input information. The complexity

TABLE 2-1   Runoff curve numbers for hydrologic soil-cover complexes[a]
(Antecedent moisture condition II; $I_a = 0.2S$)

Land use	Cover Treatment or practice	Hydrologic condition	Hydrologic soil group A	B	C	D
Fallow	Straight row	—	77	86	91	94
Row crops	Straight row	Poor	72	81	88	91
	Straight row	Good	67	78	85	89
	Contoured	Poor	70	79	84	88
	Contoured	Good	65	75	82	86
	Contoured and terraced	Poor	66	74	80	82
	Contoured and terraced	Good	62	71	78	81
Small grain	Straight row	Poor	65	76	84	88
		Good	63	75	83	87
	Contoured	Poor	63	74	82	85
		Good	61	73	81	84
	Contoured and terraced	Poor	61	72	79	82
		Good	59	70	78	81
Close-seeded	Straight row	Poor	66	77	85	89
legumes[b]	Straight row	Good	58	72	81	85
or	Contoured	Poor	64	75	83	85
rotation	Contoured	Good	55	69	78	83
meadow	Contoured and terraced	Poor	63	73	80	83
	Contoured and terraced	Good	51	67	76	80
Pasture		Poor	68	79	86	89
or range		Fair	49	69	79	84
		Good	39	61	74	80
	Contoured	Poor	47	67	81	88
	Contoured	Fair	25	59	75	83
	Contoured	Good	6	35	70	79
Meadow		Good	30	58	71	78
Woods		Poor	45	66	77	83
		Fair	36	60	73	79
		Good	25	55	70	77
Farmsteads		—	59	74	82	86
Roads (dirt)[c]		—	72	82	87	89
(hard surface)[c]		—	74	84	90	92

[a] After US SCS [19].
[b] Close-drilled or broadcast.
[c] Including right-of-way.

TABLE 2-2 Runoff curve numbers for urban and suburban areas[a]
(Antecedent moisture conditions II & III; $I_a = 0.2S$)

		Curve numbers by antecedent moisture conditions							
		II				III			
Zoning classification	Percent imperviousness	Hydrologic soil groups				Hydrologic soil groups			
		A	B	C	D	A	B	C	D
Business, industrial, or commercial	75	82	88	90	91	92	95	96	97
Apartment houses	65	78	85	88	90	90	94	95	96
Schools	45	68	78	84	87	84	90	93	95
Urban residential (Lots $\pm 10{,}000$ ft^2)	40	65	77	83	86	82	89	93	94
Suburban residential (Lots $\pm 12{,}000$ ft^2)	35	62	76	82	85	80	89	92	94
Suburban residential (Lots $\pm 17{,}000$ ft^2)	30	60	74	81	84	78	88	92	93
Suburban residential	25	58	72	80	84	77	86	91	93
Parks and cemeteries	20	55	71	79	83	74	86	91	93
Unimproved areas	15	53	70	78	82	73	85	90	92
Lawns	0	45	65	75	80	66	82	88	91

[a]SCS National Engineering Handbook, Section 4, Table 15.1 "Percent of Imperviousness for Various Densities of Urban Occupancy."

of the process is a function of the plant and its density, season, soil characteristics, and meteorologic conditions such as wind profile speed, humidity, ambient temperature, cloud cover, and latitude.

Evaporation occurs through the soil primarily in the liquid phase and is vaporized at the soil–air interface. Transpiration occurs as vaporization at the surface of the plant leaves after the soil water has been transported through the plants.

For simplification, it has been assumed by many that water supply to the plant and soil surface is not limiting which permits the treatment of evaporation at its potential rate. There are three basic approaches to the deterministic modeling of potential evapotranspiration: (1) the energy budget, (2) the aerodynamic approach, and (3) the combination of energy budget and aerodynamic approach. Because of the complexity of the process, many empirical evapotranspiration expressions have been presented in the literature.

Since the main concern in this book is on stormwater, and evapotranspiration is usually negligibly small during rainfall, evapotranspiration will be mostly neglected. However, conceptual models of evapotranspiration have been incorporated into two parametric models of continuous streamflow presented in Part III.

Problems

2-1. Why are interception losses usually not included in an urban storm-water model?

2-2. Define
   (1) field capacity
   (2) wilting point
Why is moisture content of soil pores constant during these conditions?

2-3. How does soil moisture move vertically under unsaturated condition?

2-4. Define
   (1) isotropic
   (2) homogeneous
   (3) anisotropic
Give examples of each.

2-5. Derive Eq. (2-17) from Eqs. (2-8) and (2-13).

2-6. Find solutions to Eq. (2-18) and their limitations.

2-7. Derive Eq. (2-24) from Eqs. (2-19)–(2-21).

2-8. Two feet of water was ponded on a soil where the water table is 6.5 ft below the surface. The porosity of the soil is 0.35 and the wetting front reached the water table in 3.2 hr. Estimate (a) the conductivity of the soil, (b) how much water remained ponded at 3.2 hr, and (c) the average infiltration rate over the infiltrating period.

2-9. For the Bogota silt loam shown in Fig. 2-11, (a) estimate the field infiltration rate curve with the Holtan–Overton and Horton models using the following data: The rain was 3.5 in./hr and lasted for 30 min, the final rate of infiltration for this soil is 0.20 in./hr, the available water capacity above the confining layer is 4.6 in., and the initial moisture was 1.3 in. at the beginning of rainfall; (b) what is the rain excess time distribution (hyetograph) and the total rain excess volume for this storm?

2-10. What is the moisture characteristic? Explain the hysterestic effect involved.

2-11. Derive Eqs. (2-41), (2-42), and (2-43) from Eq. (2-40).

References

1. Viessman, W., Jr., Harbaugh, T. E., and Knapp, J. W., "Introduction to Hydrology." Intext, New York, 1972.
2. Kittredge, J., "Forest Influences." McGraw-Hill, New York, 1948.
3. Chow, V. T., "Handbook of Applied Hydrology." McGraw-Hill, New York, 1964.
4. Linsley, R. K., Jr., Kohler, M. A., and Paulhus, J. L. H., "Applied Hydrology." McGraw-Hill, New York, 1949.
5. Hicks, W. I., A method of computing urban runoff, *Trans. Amer. Soc. Civil Engr.* **109** (1944).

6. Viessman, W., Jr., A linear model for synthesizing hydrographs for small drainage areas, paper presented at 48th Ann. Meeting Amer. Geophys. Union, Washington, D.C., April 1967.

7. Hillel, D., "Soil and Water: Physical Principles and Processes." Academic Press, New York, 1971.

8. Overton, D. E., Mathematical refinement of an infiltration equation for watershed engineering, USDA-ARS, 41–99 (1964).

9. Zaslavsky, D., and Rogowski, A. S., Hydrologic and morphologic implications of anisotropy and infiltration in soil profile development, *Soil Sci. Soc. Amer. Proc.* **33**, No. 4, 594–599 (1969).

10. Darcy, Henry, "Les Fontaines Publique de la Ville de Dijon." Dalmont, Paris, 1856.

11. Childs, E. C., The use of soil moisture characteristics in soil studies, *Soil Sci.* **50**, 239–252 (1940).

12. Visser, W. C., Crop growth and availability of moisture, Inst. of Land and Water Mngt., Wageningen, Netherlands, Tech. Bull. No. 6, 1959.

13. Gardner, W. R., Hillel, D., and Benyamini, Y., Post irrigation movement of soil water: I. Redistribution, *Water Resources Res.* **6**(4), 1148–1153 (1970).

14. Eagleson, P. S., "Dynamic Hydrology." McGraw-Hill, New York, 1970.

15. Smith, D. E., and Woolhiser, D. A., Mathematical simulation of infiltrating watersheds, Hydrology Papers No. 47, Colorado State Univ., Jan. 1971.

16. Green, W. H., and Ampt, G. A., Studies on soil physics: I. Flow of air and water through soils, *J. Agr. Sci.* **4**, 1–24 (1911).

17. Horton, R. E., Approach toward a physical interpretation of infiltration capacity, *Proc. Soil Sci. Soc. Amer.* **5**, 399–417 (1939).

18. Musgrave, G. W., "How Much Rain Enters the Soil?" U. S. Dept. Agr. Yearbook, pp. 151–159, 1955.

19. US Soil Conservation Service, "National Engineering Handbook, Section 4, Hydrology." Washington, D.C., Aug. 1972.

20. Watson, K. K., "Some operating characteristics of a rapid response tensiometer system," *Water Resources Res.* **1**, No. 4, 577–586 (1965).

21. Holtan, H. N., A concept of infiltration estimates in watershed engineering, US Dept. of Agr., Agr. Res. Service, No. 41–51, Washington, D.C., 1961.

22. US Agricultural Research Service, "Moisture-tension data for selected soils on experimental watersheds," No. 41-144, Washington, D.C., Oct. 1968.

23. Huggins, L. F., and Monke, E. J., "The mathematical simulation of the hydrology of small watersheds," Tech. Rep. 1, Purdue Univ. Water Resources Center, Lafayette, Ind., Aug. 1966.

Chapter 3 | Overland and Open Channel Flow

3-1 Introduction

The second basic system component in deterministic stormwater models is the surface runoff models, i.e., overland and open channel flow. The approach usually taken is a simplified one-dimensional flow approximation. The main problem associated with deterministic surface runoff modeling is the difficulty associated with solving the equations of motion. However, in the last decade there have been significant advances in the science of surface water hydraulics which has resulted in the development of a substantial simplification of the flow equations. This simplification is called the kinematic wave approximation and it is now clearly established that the approximation can be made under almost all conditions of overland flow and for many conditions associated with stormwater flows in open channels.

The governing equations of flow and solution techniques will be presented in this chapter. Chapter 4 will be entirely devoted to the kinematic approximation including real world examples of overland, storm sewer, and streamflow problems.

3-2 The Governing Equations of Motion

The basic differential equations of one-dimensional unsteady flow have been around for centuries. They have only been recently utilized because it was not possible to solve them without a high speed digital computer. The equations were apparently first applied to the overland flow problem by Keulegan [1].

The basic assumptions made in development of these equations are that

1. flow is gradually varied, meaning that
2. slope of the plane is small;
3. streamlines are essentially straight, hence
4. the pressure distribution is approximately hydrostatic;
5. resistance to flow can be closely approximated by extrapolating formulas and resistance coefficients from normal flow, and
6. momentum carried to the fluid from lateral inflow is negligible;
7. the channel is rectangular.

A fluid element of flow is shown in Fig. 3-1, where Q is the flowrate in cubic feet per second (cfs), V the average velocity of flow in feet per second, q the lateral inflow rate in cubic feet per second per unit length of channel, y and A the depth and cross-sectional area of flow in feet and feet squared, respectively, x and t the space and time coordinates in feet and seconds, and $\propto$ the angle of inclination of the plane or channel.

The principle of conservation of mass states that

$$\text{inflow} - \text{outflow} = \text{change in volume stored} \tag{3-1}$$

Because the flow is gradually varied, the total inflow into the section is

$$\text{inflow} = Q - (\partial Q/\partial x)(\Delta x/2) + q\Delta x \tag{3-2}$$

Likewise, outflow from the section is

$$\text{outflow} = Q + (\partial Q/\partial x)(\Delta x/2) \tag{3-3}$$

The change in volume stored in the section would simply be equal to the change in average depth of flow times the length of the section.

$$\text{change in volume stored} = \partial A/\partial t \, \Delta x \tag{3-4}$$

Now, combining Eq. (3-2)–(3-4) into Eq. (3-1), we find

$$Q - (\partial Q/\partial x)\,(\Delta x/2) + q\,\Delta x - Q - (\partial Q/\partial x)\,(\Delta x/2) = \partial A/\partial t \, \Delta x \tag{3-5}$$

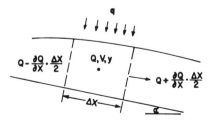

FIG. 3-1. Derivation of conservation of mass.

Upon combining terms, dividing by Δx and rearranging, the final form of the conservation of mass equation is attained

$$(\partial Q/\partial x) + (\partial A/\partial t) = q \qquad (3\text{-}6)$$

Since there are two unknowns in Eq. (3-6), there is a need for another equation. The second equation can be derived from Newton's laws of motion. The sum of all external forces acting on the water is set equal to the change in linear momentum. The external forces acting on the water body are shown on the flow profile in Fig. 3-2. These external forces are pressure gradient, weight component, and the resistive force.

As with the development of the conservation of mass equation, a pressure force p in pounds, exists at the center of the section Δx. The pressure downslope acts opposite to the pressure upslope and upon summing, the pressure gradient becomes

$$p - \tfrac{1}{2}(\partial p/\partial x)\,\Delta x - p - \tfrac{1}{2}(\partial p/\partial x)\,\Delta x = -(\partial p/\partial x)\,\Delta x \qquad (3\text{-}7)$$

where forces in the downslope direction are considered positive.

The component of the total weight of the volume of water in the section is

$$W_x = (\gamma A \sin \alpha)\,\Delta x \qquad (3\text{-}8)$$

where W_x is the weight component in the x direction per unit width, in pounds per foot, and γ is the specific weight of the fluid in pounds per cubic foot.

The resistive force is

$$R = \tau P\,\Delta x \qquad (3\text{-}9)$$

where τ is the resistive force per unit length in pounds per foot and P is wetted perimeter in feet. It is assumed that the momentum carried to the fluid by the rain is negligible. Therefore, the change in linear momentum ΔM becomes the sum of Eqs. (3-7)–(3-9):

$$\Delta M = -(\partial p/\partial x)\,\Delta x + (\gamma A \sin \alpha)\,\Delta x - \tau P\,\Delta x \qquad (3\text{-}10)$$

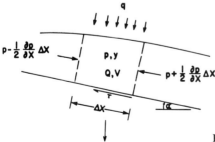

FIG. 3-2. Derivation of momentum equation.

The change in momentum consists of two parts, the local or temporal momentum change and the convective or spatial momentum change. The local momentum of the fluid is $\rho A V \Delta x$, where ρ is fluid density, and the local change is

$$(\partial/\partial t)(\rho A V \Delta x) = \rho \Delta x[A(\partial V/\partial t) + V(\partial A/\partial t)] \qquad (3\text{-}11)$$

The momentum flowing into the volume is $\rho Q V$ or $\rho A V^2$; therefore, the spatial change in momentum is

$$(\partial/\partial x)(\rho A V^2) \Delta x = \rho \Delta x[V^2(\partial A/\partial x) + 2VA(\partial V/\partial x)] \qquad (3\text{-}12)$$

and the total momentum change is the sum of Eqs. (3-11) and (3-12)

$$\Delta M = \rho \Delta x[A(\partial V/\partial t) + V(\partial A/\partial t) + V^2(\partial A/\partial x) + 2VA(\partial V/\partial x)] \quad (3\text{-}13)$$

Equation (3-13) can be simplified since from the continuity equation Eq. (3-6),

$$V(\partial A/\partial x) + A(\partial V/\partial x) = q - (\partial A/\partial t) \qquad (3\text{-}14)$$

and upon substituting Eq. (3-14) into Eq. (3-12),

$$\Delta M = \rho \Delta x[A(\partial V/\partial t) + AV(\partial V/\partial x) + Vq] \qquad (3\text{-}15)$$

By equating (3-10) and (3-15), the equation of motion becomes

$$\rho \Delta x[A(\partial V/\partial t) + AV(\partial V/\partial x) + Vq] = [-|\partial p/\partial x) + \gamma A \sin \alpha - \tau P] \Delta x \qquad (3\text{-}16)$$

Equation (3-16) can be placed in the form of

$$V(\partial V/\partial x) + (\partial V/\partial t) + g(\partial y/\partial x) = g(S_0 - S_f) - (Vq/A) \qquad (3\text{-}17)$$

where S_0 is bed slope and S_f is the friction slope and is defined as

$$S_f = \tau/\gamma R \qquad (3\text{-}18)$$

where R is hydraulic radius in feet and is equal to A/P.

Equations (3-6) and (3-17) are applicable to any cross section of both overland and open channel flow, though rigorously they apply to rectangular channels only.

The governing equations are nonlinear, hyperbolic, partial differential equations and are a nonlinear, deterministic, distributed, time-variant system representation.

3-3 Kinematic and Dynamic Waves

When the inertia and pressure forces are important in Eq. (3-17), dynamic waves govern the movement of long waves in shallow water, e.g., a flood wave in a wide river [2]. When the inertia and pressure terms are not important, kinematic waves govern flow. The weight component is essentially balanced by the resistive force indicating that the fluid is not appreciably accelerating and the flow is nearly uniform.

The analysis of Lighthill and Whitham [3] showed that when the average velocity of flow V is greater than twice the speed of the wave relative to the water, $\sqrt{gy}$, the depth of flow will continue to increase and a surge or bore will develop. This ratio of fluid speed to wave speed or celerity is called Froude number $\mathbf{F}$. They also found that when $\mathbf{F}$ was exactly equal to 2, kinematic waves prevailed over dynamic waves. Generally, the kinematic approximation would be good when $\mathbf{F}$ is greater than 1 (supercritical flow) because the waves could not move upstream since $V > \sqrt{gy}$. If there is no wave movement upstream, then flow approaches a uniform condition. Hence, kinematic flow is an unsteady uniform flow approximation.

Lighthill and Whitham [3] also found that dynamic waves are damped if

$$\mathbf{F} < 2 \tag{3-19}$$

Woolhiser and Ligget [4] found that for most practical conditions of overland flow and small watershed channel flow, dynamic waves will be small and the kinematic wave solution will closely approximate the complete momentum equation, Eq. (3-17). However, subcritical kinematic flow is both theoretically possible and has been observed. This will be explained in Chapter 4.

The kinematic approximation can be shown by operating upon Eq. (3-17) and placing it in the form

$$Q = Q_n\{1 - (1/S_0)[(\partial y/\partial x) + (V/g)(\partial V/\partial x) + (1/g)(\partial V/\partial t) + (qV/gA)]\}^{1/2}$$

$$\tag{3-20}$$

This operation can be done by extrapolating a normal flow formula such as Manning's to unsteady nonuniform flow such that

$$Q = \frac{1.49}{n} AR^{2/3} \sqrt{S_f} \tag{3-21}$$

where n is the Manning roughness coefficient.

When the dynamic terms, i.e., inertia, pressure, and local inflow are small relative to the unit weight of water, S_0, then flow is approximately uniform

$$Q \approx Q_n \tag{3-22}$$

Another way of stating the approximation is that kinematic flow will exist when discharge or flowrate is a unique function of depth of flow. For overland flow, Manning's formula reduces to

$$Q_n = \frac{1.49}{n} wy^{5/3} \sqrt{S_0} \qquad (3\text{-}23)$$

since radius is approximately equal to depth for a wide rectangular channel such as overland flow. Many experimentally determined rating curves obtained in streams by the US Geological Survey are single valued.

3-4 Solution Techniques

Method of Characteristics

The method of characteristics is a semigraphical technique whereby analytical solutions of the equations of motion can be obtained if they exist, and numerical solutions (finite differences) can be worked out. The essence of the method is to trade four partial differential equations for four ordinary differential equations.

By making the transformation

$$c^2 = gy \qquad (3\text{-}24)$$

in Eqs. (3-6) and (3-17), and then by adding and subtracting the new equations, two new equations result which are in the form of directional derivatives of $V + 2c$. The solutions of the new equations then are readily obtained by integrating along the positive and negative characteristic curves. A thorough examination of the method of characteristics and other methods was reported by Liggett and Woolhiser [5].

Combining Eq. (3-24) with Eqs. (3-6) and (3-17) and then adding and subtracting, results in

$$(V \pm c)\frac{\partial(V \pm 2c)}{\partial x} + \frac{\partial(V \pm 2c)}{\partial t} = g(S_0 - S_f) - (V \mp c)\frac{q}{A} \qquad (3\text{-}25)$$

The next step is to combine the governing equations, (3-6) and (3-17) with the total differentials of depth and velocity, which are

$$dV = (\partial V/\partial x)\,dx + (\partial V/\partial t)\,dt \qquad (3\text{-}26)$$

and

$$dy = (\partial y/\partial x)\,dx + (\partial y/\partial t)\,dt \qquad (3\text{-}27)$$

Considering the derivative terms as variables, the four equations are in linear algebraic form and can be written as

$$\begin{bmatrix} 0 & V & 1 & g \\ w & A & 0 & wV \\ dt & 0 & 0 & dx \\ 0 & dx & dt & 0 \end{bmatrix} \begin{bmatrix} \partial y/\partial t \\ \partial V/\partial x \\ \partial V/\partial t \\ \partial y/\partial x \end{bmatrix} = \begin{bmatrix} g(S_0 - S_f) - (V \pm c)(q/A) \\ q \\ dy \\ dV \end{bmatrix}$$ (3-28)

There is but a unique solution if the determinant of the matrix is non-zero, and since we are interested in an infinity of solutions, the coefficient matrix is set to zero and is expanded as a determinant. The result is

$$dx/dt = V \pm c$$ (3-29)

dx/dt is called a characteristic and is the speed of a wave relative to an observer standing on the bank. Combining Eq. (3-29) with Eq. (3-25) results in

$$\frac{\partial(V \pm 2c)}{dt} = \frac{dx}{dt} \frac{\partial(V \pm 2c)}{\partial x} + \frac{\partial(V \pm 2c)}{\partial t}$$ (3-30)

Hence, Eq. (3-25) reduces to an ordinary differential equation

$$\frac{d(V \pm 2c)}{dt} = g(S_0 - S_f) - (V \mp c)\frac{q}{A}$$ (3-31)

The four partial differential equations in equation set (3-28) are replaced with Eqs. (3-29) and (3-31).

The solution zones are represented in Fig. 3-3, taken from Woolhiser and Liggett [4]. These zones are formed by the intersection of the forward and backward characteristics. The solutions depend upon whether the zone touches the upstream $(x = 0)$ or downstream $(x = 1)$ boundaries. In zone A for example, the solution touches $x = 0$ and 1 at only one point and therefore, the solutions in this zone is not dependent on the boundary

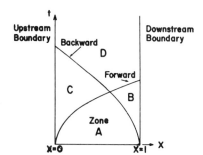

FIG. 3-3. Solution domain of hydraulic equations. (After Woolhiser and Liggett, *Water Resources Res.* **3**, No. 3, 753–771 (1967), copyright by American Geophysical Union [4].)

conditions. The solution in zone B is dependent on the downstream boundary condition; the solution in zone C is dependent on the upstream boundary condition; and the solution in zone D is dependent upon both the upstream and downstream condition. The solution in all zones is dependent on the initial conditions of the surface.

There are no known analytical solutions of the characteristic equations for streamflow, but a closed form solution was found by Woolhiser and Liggett [4] for overland flow and will be reported in Chapter 4.

The use of the method of characteristics, Eqs. (3-29) and (3-31), applies only to the $x - t$ plane shown in Fig. 3-3 and is not applicable to any other plane.

Numerical Methods

There are two basic types of numerical methods or finite-differencing techniques utilized in solving the shallow water equations or the characteristic equations. They are the explicit and implicit schemes. Miller [6] has reported a complete treatise on the subject, and the work of Amein and Fang [7] provides notable computational examples.

Explicit methods applied to either the characteristic network or to the governing equations usually result in linear algebraic equations from which the unknowns can be evaluated directly without iterative computations. Implicit methods involve nonlinear algebraic finite difference equations whereby the solution is attained by iteration. Both type methods can be and have been utilized in solving the governing equations and the characteristic netwoek. In solving the characteristic equations, either a fixed mesh or a characteristic network can be used for both method types. A fixed mesh is normally used for both method types for solving the governing equations.

Examples are shown here of solving the characteristic equations using an explicit and an implicit method with a characteristic network as well as examples of solving the governing equations using an explicit and implicit method with a fixed mesh. Both finite difference integrations of the characteristic equations result in the solution of one node point at a time. This is also true of the explicit fixed mesh integration of the governing equations, however, the implicit fixed mesh integration of the governing equations results in the simultaneous solutions of all node points. When the finite differencing technique calculates one node point at a time, the solution is said to "march out."

There are advantages and disadvantages to both approaches. Explicit methods are only conditionally stable, meaning that errors will grow as the solution progresses and are a function of the step sizes of time and distance. Trial and error computer solutions are necessary to establish stability criteria. However, explicit methods are easy to program and are generally easy to handle. Implicit methods are unconditionally stable and are more computationally efficient than explicit methods but are considerably more difficult to program.

There are several explicit and implicit numerical schemes which have been developed and utilized since the advent of the high speed digital computer. These methods are thoroughly discussed by Miller [6]. An example explicit and implicit grid network will be shown here to contrast the two approaches. The finite differencing techniques will be used to solve the characteristic equations.

The characteristic network is shown in Fig. 3-4, and explicit and implicit solutions will be shown. An explicit solution of Eqs. (3-29) for the characteristic network with a constant distance increment Δx is

$$x_p - x_A = (t_p - t_A)(V + \sqrt{gy})_A \tag{3-32}$$

and

$$x_B - x_p = (t_B - t_p)(V - \sqrt{gy})_B \tag{3-33}$$

Values of the hydraulic variables at node points A and B are specified by the initial and boundary values. It remains only to solve for x_p and t_p from these two equations. Equations (3-31) are solved by the same approach.

$$(V + 2c)_p - (V + 2c)_A = (t_p - t_A)\lambda_A \tag{3-34}$$

and

$$(V - 2c)_B - (V - 2c)_p = (t_B - t_p)\lambda_B \tag{3-35}$$

where

$$\lambda = g(S_0 - S_f) - qV/A \tag{3-36}$$

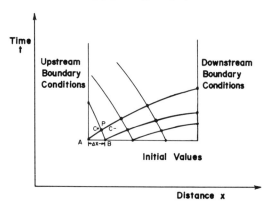

FIG. 3-4. Characteristic network.

conditions. The solution in zone B is dependent on the downstream boundary condition; the solution in zone C is dependent on the upstream boundary condition; and the solution in zone D is dependent upon both the upstream and downstream condition. The solution in all zones is dependent on the initial conditions of the surface.

There are no known analytical solutions of the characteristic equations for streamflow, but a closed form solution was found by Woolhiser and Liggett [4] for overland flow and will be reported in Chapter 4.

The use of the method of characteristics, Eqs. (3-29) and (3-31), applies only to the $x - t$ plane shown in Fig. 3-3 and is not applicable to any other plane.

Numerical Methods

There are two basic types of numerical methods or finite-differencing techniques utilized in solving the shallow water equations or the characteristic equations. They are the explicit and implicit schemes. Miller [6] has reported a complete treatise on the subject, and the work of Amein and Fang [7] provides notable computational examples.

Explicit methods applied to either the characteristic network or to the governing equations usually result in linear algebraic equations from which the unknowns can be evaluated directly without iterative computations. Implicit methods involve nonlinear algebraic finite difference equations whereby the solution is attained by iteration. Both type methods can be and have been utilized in solving the governing equations and the characteristic netwoek. In solving the characteristic equations, either a fixed mesh or a characteristic network can be used for both method types. A fixed mesh is normally used for both method types for solving the governing equations.

Examples are shown here of solving the characteristic equations using an explicit and an implicit method with a characteristic network as well as examples of solving the governing equations using an explicit and implicit method with a fixed mesh. Both finite difference integrations of the characteristic equations result in the solution of one node point at a time. This is also true of the explicit fixed mesh integration of the governing equations, however, the implicit fixed mesh integration of the governing equations results in the simultaneous solutions of all node points. When the finite differencing technique calculates one node point at a time, the solution is said to "march out."

There are advantages and disadvantages to both approaches. Explicit methods are only conditionally stable, meaning that errors will grow as the solution progresses and are a function of the step sizes of time and distance. Trial and error computer solutions are necessary to establish stability criteria. However, explicit methods are easy to program and are generally easy to handle. Implicit methods are unconditionally stable and are more computationally efficient than explicit methods but are considerably more difficult to program.

There are several explicit and implicit numerical schemes which have been developed and utilized since the advent of the high speed digital computer. These methods are thoroughly discussed by Miller [6]. An example explicit and implicit grid network will be shown here to contrast the two approaches. The finite differencing techniques will be used to solve the characteristic equations.

The characteristic network is shown in Fig. 3-4, and explicit and implicit solutions will be shown. An explicit solution of Eqs. (3-29) for the characteristic network with a constant distance increment Δx is

$$x_p - x_A = (t_p - t_A)(V + \sqrt{gy})_A \qquad (3\text{-}32)$$

and

$$x_B - x_p = (t_B - t_p)(V - \sqrt{gy})_B \qquad (3\text{-}33)$$

Values of the hydraulic variables at node points A and B are specified by the initial and boundary values. It remains only to solve for x_p and t_p from these two equations. Equations (3-31) are solved by the same approach.

$$(V + 2c)_p - (V + 2c)_A = (t_p - t_A)\lambda_A \qquad (3\text{-}34)$$

and

$$(V - 2c)_B - (V - 2c)_p = (t_B - t_p)\lambda_B \qquad (3\text{-}35)$$

where

$$\lambda = g(S_0 - S_f) - qV/A \qquad (3\text{-}36)$$

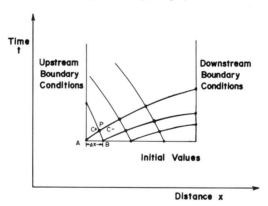

FIG. 3-4. Characteristic network.

Equations (3-34) and (3-35) are linear and V_p and c_p can be easily solved. The upstream boundary conditions (inflow hydrograph) and initial values (state of system at $t = 0$) must be known in order to start the solution. A downstream boundary condition (rating between discharge and depth) must also be specified. Comprehensive treatment of the upstream and downstream boundary condition is reported by Amein and Fang [7].

The nature of characteristics is such that the wave trains in the $x - t$ plane are usually not orthogonal. Hence, it would be fortuitous for a fixed mesh to accurately document the characteristic network. This is why Δx and Δt must be so very small in explicit solutions to achieve an acceptable level of stability.

Referring to Fig. 3-4, the space coordinate x is again held constant and the implicit difference equations for the first node point p are

$$x_p - x_A = (t_p - t_A)\tfrac{1}{2}[(V + c)_p + (V + c)_A] \quad (3\text{-}37)$$

$$x_B - x_p = (t_B - t_p)\tfrac{1}{2}[(V - c)_B + (V - c)_p] \quad (3\text{-}38)$$

$$(V + 2c)_p - (V + 2c)_A = \tfrac{1}{2}(t_p - t_A)(\lambda_p + \lambda_A) \quad (3\text{-}39)$$

and

$$(V - 2c)_B - (V - 2c)_p = \tfrac{1}{2}(t_B - t_p)(\lambda_B + \lambda_p) \quad (3\text{-}40)$$

In the four difference equations, the unknowns are x_p, t_p, V_p, y_p, and S_{fp}. However, V_p, y_p, and S_{fp} are interrelated and therefore there are only four unknowns. The finite difference equations (3-37) through (3-40) are algebraically nonlinear and must be solved using an iterative numerical method. Amein and Fang [7] reported much success in utilizing Newton's iteration method in the solution of these equations.

The explicit and implicit differencing schemes can also be used to solve the governing equations as an alternative to the solution of the characteristic equations. The application of the explicit method to the unsteady flow equations is primarily the outcome of the pioneering work of J. J. Stoker. A complete description of the numerical solutions of the governing equations of unsteady flow is given by Issacson et al. [8]. The explicit scheme shown here is from that report.

A network of node points is shown in Fig. 3-5 for solving the governing equations using the explicit method. The centered difference solution scheme is used to solve for the hydraulic variables at node point M, but by approximating the partial derivatives with finite differences, the solution at point P is had. In Eqs. (3-6) and (3-17), the following approximations are made at node point M:

$$V(M) = [V(R) + V(L)]/2 \quad (3\text{-}41)$$

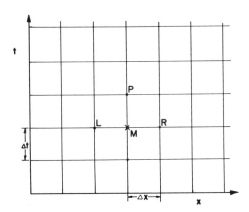

FIG. 3-5. Network of points for solving the unsteady flow equations using the explicit method. (After Issacson *et al.* [8].)

$$\partial A(M)/\partial t = [A(P) - A(M)]/\Delta t \qquad (3\text{-}42)$$

$$\partial V(M)/\partial x = [V(R) - V(L)]/2\Delta x \qquad (3\text{-}43)$$

$$\partial A(M)/\partial x = [A(R) - A(L)]/2\Delta x \qquad (3\text{-}44)$$

$$\partial V(M)/\partial t = [V(P) - V(M)]/\Delta t \qquad (3\text{-}45)$$

$$\partial y(M)/\partial x = [y(R) - y(L)]/2\Delta x \qquad (3\text{-}46)$$

$$S_f(M) = [S_f(R) + S_f(L)]/2 \qquad (3\text{-}47)$$

$$\partial Q(M)/\partial x = [Q(R) - Q(L)]/2\Delta x \qquad (3\text{-}48)$$

$$q(M) = [q(R) + q(L)]/2 \qquad (3\text{-}49)$$

When these approximations are inserted into Eqs. (3-6) and (3-17), $V(P)$ and $y(P)$ can be solved for directly as

$$\begin{aligned} V(P) = \tfrac{1}{2}[V(R) + V(L)] &- (\Delta t/\Delta x)\{\tfrac{1}{2}[V(R)^2 - V(L)^2] \\ &+ g[V(R) + q(R) - y(L) - q(L)] \\ &+ g\{\tfrac{1}{2}[S_f(R) + S_f(L)]\}\,\Delta x\} \end{aligned} \qquad (3\text{-}50)$$

$$y(P) = \tfrac{1}{2}\{y(R) + y(L)\} + (1/w)(\Delta t/\Delta x)\{A(L)\cdot V(L) - A(R)\cdot V(R)\} \qquad (3\text{-}51)$$

The stability of the computations is determined by the ratio of the grid sizes $\Delta t/\Delta x$, and a necessary but insufficient condition for stability is

$$\Delta t \le \Delta x/|V + c| \qquad (3\text{-}52)$$

This criterion for step sizes, known as the Courant stability condition implies that the time increment Δt must be selected such that the node point P will lie within the area bounded by the characteristics generated from node points L and R.

This scheme has been utilized by the Tennessee Valley Authority [9] in their unsteady river computational system.

A network of node points is shown in Fig. 3-6 for solving the unsteady flow equations using the implicit method. Rather than determine the values of the hydraulic variables at a future time step by projecting from known values from a previous time step as in the explicit method, relations existing among the hydraulic variables at the future time step are used in the implicit method. The central difference implicit scheme reported by Brakenseik et al. [10] and Amein and Fang [7] will be demonstrated here.

The following approximations are made for the hydraulic variables at point M:

$$\left.\frac{\partial Q}{\partial x}\right|_M = \frac{Q(4) + Q(3) - Q(2) - Q(1)}{2\,\Delta x} \tag{3-53}$$

$$\left.\frac{\partial A}{\partial t}\right|_M = \frac{A(4) + A(2) - A(3) - A(1)}{2\,\Delta t} \tag{3-54}$$

$$\left.\frac{\partial y}{\partial x}\right|_M = \frac{y(4) + y(3) - y(2) - y(1)}{2\,\Delta x} \tag{3-55}$$

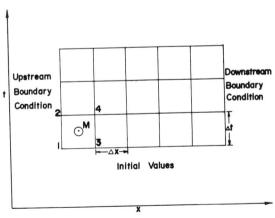

FIG. 3-6. Network of points for solving the unsteady flow equations using the implicit method.

$$\frac{\partial V}{\partial x}\bigg|_M = \frac{V(4) + V(3) - V(2) - V(1)}{2\,\Delta x} \tag{3-56}$$

$$\frac{\partial V}{\partial t}\bigg|_M = \frac{V(4) + V(2) - V(3) - V(1)}{2\,\Delta t} \tag{3-57}$$

$$S_f(M) = \tfrac{1}{4}[S_f(1) + S_f(2) + S_f(3) + S_f(4)] \tag{3-58}$$

$$q(M) = \tfrac{1}{4}[q(1) + q(2) + q(3) + q(4)] \tag{3-59}$$

These approximations are then inserted into Eqs. (3-6) and (3-17). Hydraulic variables at node points 1, 2, and 3 are known from boundary conditions and initial values, hence the unknowns are $Q(4)$, $V(4)$, $y(4)$, $A(4)$, and $S_f(4)$. Since $y(4)$ and $A(4)$ are related and $Q(4)$, $V(4)$, and $A(4)$ are related, there are actually three unknowns and two equations. Since there is a need for another equation, the difference scheme is written for all of the time steps in the problem until the downstream boundary condition is reached. In Fig. 3-6, there are 12 grid boxes, meaning that there will be 24 equations to be written but there will be 27 unknowns. The three additional equations are specified by the downstream boundary condition which will be a rating between discharge and area of the form

$$Q = aA^b \tag{3-60}$$

where a and b are experimentally determined.

Amein and Fang [7] found that the Newton iteration scheme was also very effective in solving this set of nonlinear algebraic equations.

The theory and solution techniques presented in this chapter establish a basis for deterministic surface water modeling and it will be utilized in the forthcoming chapters in model development and application.

Problems

3-1. Derive Eq. (3-17) from Eq. (3-16).

3-2. Derive Eq. (3-20).

3-3. Prove that hydraulic radius is approximately equal to depth for a wide rectangular channel.

3-4. Derive Eq. (3-25).

3-5. Derive Eq. set (3-28).

3-6. Derive Eq. (3-29).

3-7. Derive Eqs. (3-50) and (3-51).

3-8. A stream 100 ft wide and approximately rectangular has the following characteristics:

Bed slope $S_0 = 0.005$ ft/ft
Width $w = 100$ ft
Manning n-value $= 0.04$

The reach lengths are 1000 ft long and the time interval is 1 hr. Using the implicit numerical scheme for integrating conservation of mass and momentum, develop a solution which will be complete enough to solve on the computer. Develop your algorithm such that your next step would be to write the computer program.

To simplify the problem, assume that the convective and temporal velocity accelerations are negligibly small in the momentum equation. Also, assume that local inflow is zero. Use the following grid notation:

Distance, x

	9	10	11	12
time, t	5	6	7	8
	1	2	3	4

Initial conditions (values)

$Q(1) = 400$ cfs $y(1) = 1.5$ ft
$Q(2) = 385$ cfs $y(2) = 1.4$ ft
$Q(3) = 370$ cfs $y(3) = 1.2$ ft
$Q(4) = 353$ cfs $y(4) = 1.1$ ft

Boundary conditions
$Q(5) = 500$ cfs, $Q(9) = 650$ cfs upstream
$Q = 300\, y^{1.7}$ cfs downstream; node points 4, 8, and 12

Solve for streamflow discharge at node points 6, 7, 8, 10, 11, and 12.

3-9. You are observing the rising stage of a major flood in a large river; at your observation point it is seen that at a certain time the discharge is 75,000 cfs, and that the water level is rising at the rate of 1 ft/hr. The surface width of the river at this point, and for some miles upstream and downstream, is one-half mile. You are asked to make a quick estimate of the present magnitude of the discharge at a point 5 miles upstream. What is your estimate?

3-10. Define as cogently as possible "gradually varied flow."

3-11. Assuming normal flow, estimate the discharge in a triangular drainage ditch with side slopes of 1:2 at 5 ft depth if the ditch is made of concrete and sloped at 0.001.

3-12. Estimate the average friction slope over a 1000 ft reach of an approximately rectangular channel with the following data:

$$y_1 = 6 \text{ ft}, \qquad y_2 = 5.75 \text{ ft}, \qquad V_1 = 2 \text{ fps}, \qquad V_2 = 1.8 \text{ fps}$$

$$w = 100 \text{ ft}, \qquad \text{channel sloped at } 3°$$

Assume steady flow.

3-13. At a point in a rectangular stream, the following information was obtained:

$$\partial y/\partial x = 0.002 \text{ ft/ft}, \qquad S_0 = 0.003 \text{ ft/ft}, \qquad S_f = 0.002 \text{ ft/ft},$$

$$w = 100 \text{ ft}, \qquad y = 5 \text{ ft}$$

Estimate the Froude number at this point if the flow is steady.

3-14. At a point in a rectangular stream the discharge is 530 cfs associated with a depth of flow equal to 4 ft. The stream is 100 ft wide. Estimate the speed of the wave relative to an observer standing on the bank.

3-15. Estimate the discharge in a circular concrete culvert flowing full under hydrostatic pressure if $D = 36$ in. and $S_0 = 0.01$.

3-16. Define and illustrate mathematically the following finite-difference approximations to a partial derivative:
  (a) forward difference
  (b) backward difference
  (c) central difference
Define all symbols used.

3-17. Compare the error involved in each of the difference approximations of Problem 16 through use of a series representation of a typical variable (such as average velocity U).

3-18. What is meant by an explicit finite difference formulation? An implicit formulation? Use a network of points in a time-distance grid to illustrate each of these two basic types of numerical computational procedures.

3-19. Derive by any method the one-dimensional gradually varied unsteady flow equations of continuity and momentum for an open channel of arbitrary cross section. State clearly all assumptions which are made; define all symbols used; and present sketches of the control volume used in the derivation. Show the formulation of each force acting on the control volume and all mathematical details. Allow for lateral inflow, but neglect rainfall, groundwater flow, wind action, and Coriolis forces.

References

1. Keulegan, G. H., Spatially variable discharge over a sloping plane, *Trans. Amer. Geophys. Union,* Part VI, 956–958 (1944).
2. Stoker, J. J., "Water Waves." Wiley (Interscience), New York, 1957.
3. Lighthill, M. J., and Whitham, G. B., Kinematic Waves, 1, *Proc. Roy. Soc. London A* **229,** 281–316 (1955).
4. Woolhiser, D. A., and Liggett, J. A., Unsteady one-dimensional flow over a plane—the rising hydrograph, *Water Resources Res.* **3,** No. 3, 753–771 (1967).

5. Liggett, J. A., and Woolhiser, D. A., Difference solutions of the shallow-water equations, *J. Eng. Mech. Div., Amer. Soc. Civil Eng.* **9B**(EM2), 39–71 (1967).

6. Miller, W. A., Jr. Numerical solution of the equations for unsteady open-channel flow, School of Civil Eng., Georgia Inst. of Technology, Atlanta, Sept. 1971.

7. Amein, M., and Fang, C. S., Streamflow routing—with applications to North Carolina rivers, Water Resources Res. Inst. Univ. of North Carolina, Rep. No. 17, Jan. 1969.

8. Issacson, E., Stoker, J. J., and Troesch, B. A., Numerical solution of flood prediction and river regulation problems, Inst. Math. Sci. Rep. No. IMM-235, New York Univ., New York, 1956.

9. Price, J. T., Garrison, J. M., and Granju, J-P., Seminar on hydraulic transients in open channels, Tennessee Valley Authority, Knoxville, Tennessee, Feb. 1968.

10. Brakenseik, D. L., Heath, A. L., and Comer, G. H., Numerical techniques for small watershed flood routing, ARS 41-113, US Dept. of Agr., Agr. Res. Service, Washington, D.C., Feb. 1966.

Chapter 4 | Kinematic Flow Approximation

4-1 Introduction

The kinematic flow approximation has proven to be a very useful tool in stormwater modeling. In this chapter, kinematic wave theory, the state of the art, and numerical examples of real world problems will be presented for both overland and open channel flow.

4-2 The Kinematic Approximation to Overland Flow

Considerable effort has been expended on the theoretical aspects of the applicability of the flow equations to overland flow. Kinematic waves occur when the dynamic terms in the momentum equation are negligible. There is no appreciable backwater effect and discharge would be only a function of depth of flow at all x and t.

$$Q = ay^m \tag{4-1}$$

This conclusion can be understood by using the approach of Henderson [1]. The momentum equation was normalized by a steady uniform discharge Q_n.

Remember Eq. (3-20); then

$$Q = Q_n \left[1 - \frac{1}{S_0} \left(\frac{\partial y}{\partial x} + \frac{V}{g} \frac{\partial V}{\partial x} + \frac{1}{g} + \frac{\partial V}{\partial t} + \frac{qV}{gA} \right) \right]^{1/2}$$ (3-20)

If the sum of the terms to the right of the minus sign is much less than one,

$$Q \approx Q_n$$ (4-2)

This means that unsteady flows may be approximated by a series of normal flows. Turbulent normal flows are often expressed by the Manning formula.

Woolhiser and Liggett [2] reported an analytical solution of the overland flow problem for zone A in Fig. 3-3 of the characteristic plane for an initially dry surface, neglecting the momentum of the rain. The slope of the energy gradient was approximated using the Chezy formula

$$S_f = \frac{V^2}{C^2 y}$$ (4-3)

where C is the Chezy resistance coefficient.

In the solution, the resistance coefficient was assumed constant. It was assumed that the C from steady-uniform flow was a good approximation of the C for unsteady flow. The solution is an infinite series which can be closely approximated by

$$Q_* = (t_*/k)[(k^2 t_* + 1)^{1/2} - 1]$$ (4-4)

where Q_* is discharge normalized by the rain rate, t_* the time normalized by the time to equilibrium t_e, and k the dimensionless parameter, *kinematic flow number*,

$$k = \frac{S_0 L}{H_0 \mathbf{F}_0{}^2}$$ (4-5)

where L is the length of the plane, H_0 the depth of flow at the end of the plane at equilibrium, and $\mathbf{F}_0$ the Froude number at the end of the plane at equilibrium.

For large values of k, Eq. (4-4) reduces to

$$Q_* = t_*{}^{3/2}$$ (4-6)

This is the same as the kinematic wave solution obtained by Henderson and Wooding [3]. In the momentum equation (3-17), all terms are negligible relative to bed slope and therefore

$$S_0 = S_f$$ (4-7)

In terms of the Chezy relation

$$Q = Cwy^{3/2}\sqrt{S_0} \qquad (4\text{-}8)$$

Equation (4-8) is the substitute for the momentum equation, and can be combined with the conservation of mass equation Eq. (3-6) to form

$$[(\tfrac{3}{2})C\sqrt{S_0}\,y^{1/2}](\partial y/\partial x) + (\partial y/\partial t) = q/w \qquad (4\text{-}9)$$

where q/w is rainfall excess rate. Equation (4-9) is now in the form of a directional derivative

$$dy/dt = (dx/dt)(\partial y/\partial x) + (\partial y/\partial t) = q/w \qquad (4\text{-}10)$$

Then

$$dx/dt = \tfrac{3}{2}Cy^{1/2}(S_0)^{1/2} \qquad (4\text{-}11)$$

and

$$y = (q/w)t \qquad (4\text{-}12)$$

Eliminating y from Eqs. (4-8) and (4-12) results in the kinematic wave solution for the rising hydrograph

$$Q = Cw(S_0)^{1/2}(qt/w)^{3/2} \qquad (4\text{-}13)$$

Woolhiser and Liggett [2] presented solutions for the rising hydrograph for a wide range of k-values. The solutions were started using the analytic solution in zone A and the computations were completed using the characteristic method described by Liggett and Woolhiser [4]. Their results are summarized in Fig. 4-1. For a k-value of 10, about a 10% error would be induced by deleting the dynamic terms from the solution. As k increases, the error rapidly decreases. This solution has a constant Chezy-C for all flows. The kinematic solution is associated with a k-value approaching infinity.

The kinematic wave solution is completely generalized as

$$Q = a(qt/w)^m \qquad (4\text{-}14)$$

If the resistance coefficient varies with depth of flow, then the coefficient in Eq. (4-14) would be some power function of depth; but the algebraic form of Eq. (4-14) would remain the same. In the transition from laminar to turbulent flow, the coefficient in Eq. (4-14) would also be related to depth of flow by a power function. Therefore, Eq. (4-14) should be completely general for any flow condition on the rising hydrograph.

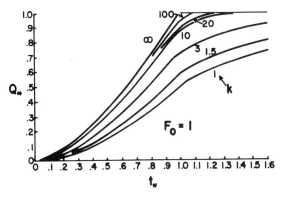

FIG. 4-1. Hydraulic solution of rising hydrograph. (After Woolhiser and Liggett, *Water Resource Res.* **3**, No. 3, 753–771 (1967), copyright by American Geophysical Union [2].)

Low flows would conceivably be laminar; but the location of the transitional Reynolds number is indeterminate. As a result of analyzing hundreds of overland flow hydrographs, the Corps of Engineers [5] have argued that raindrop impact created turbulent flow except perhaps at very low flows. The results of the analysis of the same rising hydrographs by Overton [6] supported the argument in that the observed rising hydrographs initially rose very slowly indicating a viscous flow, and then transcended to turbulent flow. But this transition cannot be explained by Reynolds number alone because the transition Reynolds number is different for each hydrograph. To illustrate, consider the extremes. The Reynolds number at equilibrium is

$$R_e = (qL/vw) \qquad (4-15)$$

where v is kinematic viscosity. The range of Reynolds number at equilibrium for the Corps' data [5] was 100 and 10,000. The transition from laminar to turbulent flow occurred at about 0.2 of the equilibrium rate for all hydrographs studied, hence the critical Reynolds numbers ranged from 20 to 2,000, respectively. It follows that rainfall intensity does have a significant effect upon the transitions. Schreiber and Bender [7] and Shen and Li [8] have studied this effect on very short planes and reached essentially the same conclusions.

4-3 Kinematic Flow Number

The kinematic wave solution for a plane using the Manning resistance relation is

$$Q_* = t_*^{5/3} \qquad (4-16)$$

Equation (4-16) is applicable to zone A only, however as shown in Fig. 4-1, zone A constitutes substantially all of the solution for kinematic flow numbers of 10 or greater.

Kinematic flow number can be placed in terms of the physical and hydraulic characteristics of a plane by eliminating H_0 and $\mathbf{F}_0$ from Eq. (4-5) using Eq. (3-23) and the equation of continuity,

$$qL = V_0 H_0 w \tag{4-17}$$

and the result is

$$k = \frac{gn^{1.2} S_0^{0.4} L^{0.2}}{1.49(q/w)^{0.8}} \tag{4-18}$$

Upon transforming the rain rate from cubic feet per second per foot into inches per hour, i, Eq. (4-18) becomes

$$k = 10^5 \frac{n^{1.2} S_0^{0.4} L^{0.2}}{i^{0.8}} \tag{4-19}$$

In general, high k-values are produced on rough, steep, long planes with low rain rates.

The laminar kinematic solution is

$$Q_* = t_*^{3} \tag{4-20}$$

and kinematic flow number is found by using the velocity law for laminar flow at the end of the plane

$$V_0 = (gS_0/3v)H_0^{2} \tag{4-21}$$

and the continuity equation (4-17) to form

$$k = 1500(gS_0/L)^{1/3} i^{-4/3} \tag{4-22}$$

4-4 Kinematic Flow on Long Impermeable Planes— the Rising Hydrograph

Attempts have been made to determine the transition from laminar to turbulent flow. One approach has been to consider the Darcy–Weisbach resistance coefficient as a function of Reynolds number [9, 10]. A modification of this approach [7] has been to apply a rainfall intensity factor to Reynolds number. The numerical value of the coefficient was determined by optimization through operation on measured laboratory hydrographs.

A different approach to the problem was reported by Overton [11] whereby Eq. (4-14) for laminar and turbulent flow was superimposed, in dimensionless form, on the rising portion of the 214 equilibrium hydrographs measured by the Corps of Engineers [5] on planes up to 500 ft long.

Equation (4-14) was formed dimensionless in order to facilitate the evaluation. At time to equilibrium t_e, discharge equals qL. Inserting these values into Eq. (4-14) and solving for time to equilibrium results in

$$t_e = (Lq^{1-m}/aw)^{1/m} \qquad (4\text{-}23)$$

If both sides of Eq. (4-14) are now divided by Eq. (4-23), the result is

$$Q_* = t_*^m \qquad (4\text{-}24)$$

where $Q_* = Q/qL$ and $t_* = t/t_e$. Use of Eq. (4-24) permits an efficient means of evaluating the kinematic wave solutions on all data. Since the observed hydrographs are generated from a wide range of rain rates, scale effects in evaluation were removed by normalizing discharge by equilibrium discharge and time by time to equilibrium [2]. Unfortunately, actual time to equilibrium is a matter of degree because transient discharge approaches equilibrium discharge asymptotically. Overton [11] found that the shape of the dimensionless rising hydrographs were very sensitive to small changes in the choice of time to equilibrium. Also, it was most difficult to determine objectively the time when flows reach 97, 98, or 99% of equilibrium. Hence, within a small variation, one could easily choose an observed time to equilibrium which would match Eq. (4-24) for any exponent value between 3/2 and 3. It was concluded that Eq. (4-24) could not be tested objectively with an arbitrary and subjective definition of time to equilibrium. Therefore, a systematic method was developed for determining time to equilibrium.

Overton [12] found an explicit mathematical relation from the kinematic equations between hydrologic lagtime and time to equilibrium. The hydrologic significance of lagtime can be seen by examining Fig. 4-2. On this

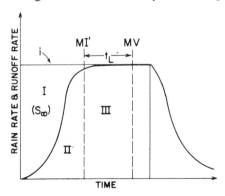

FIG. 4-2. Derivation of hydrologic lagtime. (After Overton, *Water Resource Res.* **6**, No. 1, 43–52 (1970), copyright by American Geophysical Union [2].)

conceptual equilibrium hydrograph, MI' and MV denote times of occurrence of 50% of rainfall and 50% of runoff volume. From geometry it can be shown that area I is equal to area III. But area I is the total storage on the plane at equilibrium S_∞. Thus,

$$S_\infty = qt_L/w \tag{4-25}$$

where t_L is lagtime. Actually, lagtime could be the time lapse between any given percentage of rainfall or runoff volume as long as the associated discharges are at equilibrium. Throughout equilibrium, a volume of water equal to S_∞ drains from the plane in a time equal to t_L.

A solution for lagtime was obtained from the kinematic equations. At equilibrium, Eq. (3-6) becomes

$$Q = qx \tag{4-26}$$

A solution for S_∞ was found by combining Eq. (4-14) with Eq. (4-26) and integrating

$$S_\infty = \left(\frac{m}{m+1}\right)\left(\frac{qL}{a}\right)^{1/m} \tag{4-27}$$

From Eq. (4-25), it follows that

$$t_L = \left(\frac{m}{m+1}\right)\left(\frac{Lq^{1-m}}{a}\right)^{1/m} \tag{4-28}$$

By comparing Eq. (4-23) and (4-28), it is apparent that lagtime is a factor of time to equilibrium,

$$t_e = \left(\frac{m+1}{m}\right)t_L \tag{4-29}$$

Lagtime can be determined systematically and with little error. Each of the kinematic models was evaluated in terms of lagtime. Upon substituting Eq. (4-29) into Eq. (4-24), a generalized solution of the rising hydrograph is found to be

$$Q_* = \left(\frac{m}{m+1}\right)^m \left(\frac{t}{t_L}\right)^m \tag{4-30}$$

Lagtime was determined from each observed hydrograph. Discharges were predicted for each of the three kinematic models by using the appropriate exponent in Eq. (4-30).

A large number of overland flow hydrographs collected for airfield drainage investigations has been reported by the Corps of Engineers [5].

The experiments were conducted in three concrete troughs each 500 ft long. Flows were developed from rainfall simulators over the entire surface, flows at the end of the troughs were measured by ogee-weirs and depth measurements were made by a manometer setup at three equally spaced points along the trough. The three troughs were sloped at $\frac{1}{2}$, 1, and 2% and artificial roughnesses were generated by placing expanded metal plates, excelsior pads, and chicken wire in the troughs. An unsuccessful attempt was made to grow grass in one of the troughs.

Because of the difficulty in growing grass uniformly, artificial surfaces were used to roughen the concrete. The Waterways experiment station at Vicksburg developed a similarity criteria based upon Manning n-values. The substituted material was formed from #2.5 expanded metal in 3 ft squares and placed on top of the concrete. This cover was designated as simulated turf. Chicken wire was added to make the surface rougher and excelsior pads were used to simulate dense grass.

There were 601 runoff hydrographs measured. Hydrographs were developed over lengths of 84, 168, 252, 336, 420, and 500 ft. Rainfall rates were varied from 0.25 to 10 in./hr. In the previous investigations of Horton [13] and Izzard [14], the product of maximum rain rate times plot length was about 500. In the data collected by the Corps [5], this product was 5000.

Of the hydrographs reported, 403 were developed on the concrete surface and 98 were developed on the simulated turf surface. Of these 501 runoff hydrographs, 287 were developed from various combinations of unsteady rain on nonuniform surfaces. The remaining 214 hydrographs were developed from long, steady uniform rain intensities over uniform concrete or simulated turf surfaces. These 214 events were called equilibrium hydrographs.

In the study by Overton [11], the rising portion of all 214 hydrographs were normalized. Discharge scale was normalized by the rain rate, and the time scale was normalized by the associated lagtime t_L. All normalized hydrographs were plotted on transparent paper and superimposed. Hydrographs were also grouped by slope and by cover. It was apparent that within a small error a single dimensionless rising hydrograph would accurately represent all 214 hydrographs. This is essentially what the Corps [5] found when the time scale was normalized by a time to equilibrium.

The same dimensionless rising hydrograph resulted within very small errors for all equilibrium runs. Standard deviations were very low at high flows, but even at low flows, the absolute value of the standard deviation was only ± 0.02. All variations due to slope and surface roughness were insignificant and the average rising hydrograph was used to represent all 214 events. Average hydrographs for all 214 hydrographs and for all hydrographs on concrete and simulated turf are shown in Fig. 4-3. The average

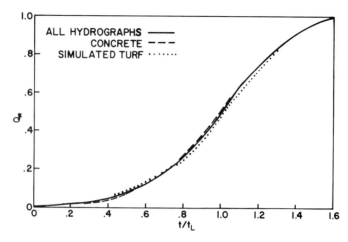

FIG. 4-3. Averaged rising hydrographs of all events for concrete and for simulated turf.

rising hydrograph will be, subsequently, referred to as the "observed."

The three kinematic models and the observed are shown in Fig. 4-4. Flows appear to be laminar during the first half of the period of rise, whereas flows appear to be fully turbulent on the upper half of the rise. This effect has been noticed by previous investigators [9, 10], however, the Corps of Engineers [5] suggested that due to raindrop impact all but the very low flows appeared to be turbulent. It is not possible to resolve this issue without velocity profiles.

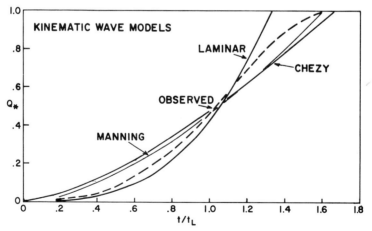

FIG. 4-4. Laminar and turbulent kinematic wave solutions and the observed rising hydrograph.

It is significant that a single dimensionless hydrograph very accurately characterizes all 214 rising hydrographs. This implies that Reynolds number alone is not the sole criteria of the viscous-turbulent transition. Since the observed in Fig. 4-4 represents equilibrium hydrographs over a range of 84 to 500 ft and rainfall rates of 0.25 to 10 in./hr, the transitional Reynolds number will vary with rainfall rate. Perhaps this is in agreement with the work of Schreiber and Bender [7] whereby a rainfall factor to be used in combination with the Reynolds number was derived from measured hydrographs.

All models appear to fit the data reasonably well. The Manning-kinematic solution produced a 15% standard error in fitting the observed, whereas both the Chezy and laminar solutions produced a 19% standard error.

The buildup of the rising hydrograph is demonstrated in Fig. 4-5. The equilibrium depth profile is denoted as $A - B_1 - B_2 - C_3$ and will reach this state at a time equal to t_e after the beginning of the steady uniform rain rate. The profiles prior to equilibrium are shown as $A - B_1 - C_1$ and $A - B_2 \, \mathbf{M} \, C_2$. In analyzing the rising hydrograph, it is important to remember that the characteristic equations (4-9) and (4-11) apply only to the characteristic $(x - t)$ plane. Hence, the depths B_1 and B_2 represent points where the wave has passed and flow has come to equilibrium at those points at times equal to

$$t_e(B_1) = x_1/V_1(B_1) \qquad (4\text{-}31)$$

and

$$t_e(B_2) = x_2/V(B_2) \qquad (4\text{-}32)$$

respectively. Flow is uniform downslope from B_1 at time equal to $t_e(B_1)$ and from B_2 at time equal to $t_e(B_2)$. Therefore, as the kinematic wave passes, flows upslope of this point are at equilibrium and will continue in this state as long as rain rate remains steady and uniform above this point.

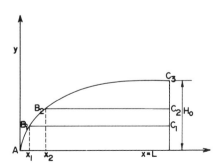

FIG. 4-5. Buildup of rising hydrograph as shown by depth hydrograph.

This is a physical explanation of why the solution in zone A is not a function of the downstream boundary condition.

4-5   Time to Equilibrium

For Manning-kinematic flow, time to equilibrium in minutes is found from Eq. (4-23) as

$$t_e = \frac{0.928}{i^{0.4}} (nL/\sqrt{S_0})^{0.6} \qquad (4-33)$$

where i is the steady uniform rain rate in inches per hour.

EXAMPLE 4.1   Estimate the time to equilibrium on a smooth concrete driveway uniformly sloped at 1% and 150 ft long for a rain rate of 1in./hr. Assume $n = 0.015$.

Using Eq. (4-33), we find

$$t_e = \frac{0.928}{(1)^{0.4}} \left(\frac{0.015(150)}{(0.01)^{1/2}} \right)^{0.6} = 6.0 \text{ min}$$

Hence, a rain intensity of 1 in./hr will bring the driveway to equilibrium in 6 min.

4-6   Equilibrium Depth Profile

The equilibrium depth profile can be estimated by solving Eqs. (4-26) and (3-23) simultaneously

$$y(x) = \left(\frac{nqx}{1.49w \sqrt{S_0}} \right)^{0.6} \qquad (4-34)$$

EXAMPLE 4.2   Estimate the equilibrium depth at the end of the concrete driveway in Example 4.1.

The rain rate in cubic feet per second per foot width per foot length is found by knowing that 1 acre-in. ≈ 1.008 cfs-hr. The conversion is

$$q/w = i/43908 = 2.28 \times 10^{-5} \text{ cfs/ft}^2$$

From Eq. (4-34),

$$H_0 = \left[\frac{0.015(2.28 \times 10^{-5})(150)}{1.49 \sqrt{0.01}} \right]^{0.6} = 8.36 \times 10^{-3} \text{ ft}$$

or $H_0 = 0.10$ in.

4-7 The Falling Hydrograph

Henderson and Wooding [3] derived the kinematic equations for the falling hydrograph. There are two cases involved in the computation of the falling hydrograph: I. when the rising hydrograph is at equilibrium, and II. when the rising hydrograph is at a flow less than equilibrium.

Case I Duration of storm, $D \geq t_e$. After rainfall stops, it can be seen from Eq. (4-10) that

$$dy/dt = 0 \tag{4-35}$$

Hence, the depth, discharge and wave speed dx/dt remain constant along a characteristic. This means that beginning with a point on the equilibrium profile and realizing that the future coordinates of that depth will lie on a single characteristic, dx/dt can be used to locate the point in space at any future time. This principle is illustrated in Fig. 4-6.

The depth profile at a time Δt after the end of rain is noted as $A - B_2 - C_2$ and the associated Δx in the characteristic plane is

$$\Delta x = amy^{m-1} \, \Delta t \tag{4-36}$$

Then the x coordinate during the recession would be

$$x = x(D) + \Delta x \tag{4-37}$$

and from Eqs. (4-34) and (4-36), Eq. (4-37) becomes

$$x = (1.49w \, \sqrt{S_0} \, y^{5/3}/nq) + (1.49/n)\tfrac{5}{3} \sqrt{S_0} \, y^{2/3}(t - D) \tag{4-38}$$

which can be factored to

$$x = (1.49/n)\sqrt{S_0} \, y^{2/3}[yw/q + \tfrac{5}{3}(t - D)] \tag{4-39}$$

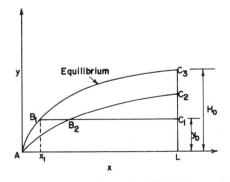

FIG. 4-6. Depth profiles on the falling hydrograph.

The depth is found from Eq. (4-39)

$$t = \tfrac{3}{5}[(nL/1.49\sqrt{S_0})y^{-2/3} - (yw/q)] + D \qquad (4\text{-}40)$$

Case II   Duration of storm, $D < t_e$. If the rain stops prior to reaching equilibrium, then the depth profile at $t = D$ would correspond to one similar to $A - B_2 - C_1$ in Fig. 4-6. Point B_1 will then move to coordinate B_2 and C_1 to coincide with that which occurs at a time t_p. This time is expressed as

$$t_p = D + \frac{L - x_1}{dx/dt} = D + \frac{t_e - D}{5/3} \qquad (4\text{-}41)$$

The discharge at the end of the plane will remain constant between $D < t < t_p$ and will be

$$Q = (1.49/n)w\sqrt{S_0}\,y_D^{5/3} \qquad (4\text{-}42)$$

The discharge hydrograph for times $> t_p$ is specified by Eq. (4-40).

EXAMPLE 4.3   For the driveway in Example 4.1, estimate the rising and falling hydrograph if the duration of the rain was 4 min.

The discharge at 4 min as found from Eq. (4-24)

$$Q = 1(4/6)^{1.67} = 0.51 \text{ in./hr}$$

From 0–4 min, the discharge can be found from Eq. (4-24). From Eq. (4-41),

$$t_p = 4 + \frac{6 - 4}{5/3} = 5.2 \text{ min}$$

The discharge then remains constant at 0.51 in./hr between 4 and 5.2 min. For times greater than 5.2 min, the recession may be calculated from Eq. (4-40) as

$$t = \tfrac{3}{5}\{15.1\,y^{-2/3} - 732y\} + t_p$$

The depth hydrograph calculated here is converted into the falling discharge hydrograph using Manning's equation. The entire hydrograph is plotted in Fig. 4.7.

4-8   Model for a V-Shaped Watershed

Wooding [15], Liggett and Woolhiser [16] and Overton and Brakensiek [17] have presented models of V-shaped watershed flow. The V-shaped

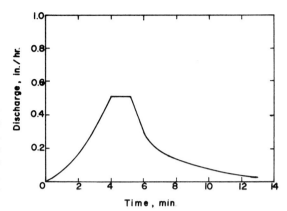

FIG. 4-7. Reprinted by permission of UNESCO and IAHS from "Results of Research on Representative and Experimental Basins," Proceedings of the Wellington Symposium, Dec. 1970, Vol. 1 © UNESCO–IAHS 1973.

watershed is formed by two planes intersecting a channel as shown in Fig. 4-8. The kinematic equations have been used to describe flow on the planes as well as in the channel whereby the lateral inflow into the channel is specified as the overland flow at the end of the planes.

As shown by Liggett and Woolhiser [16] (see Fig. 4-9), there are three different solution zones in the characteristic plane of the channel and these solution zones depend upon the relation between the time t_1 that the first characteristic in the channel reaches L_c relative to the time to equilibrium

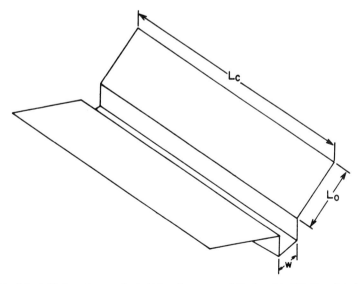

FIG. 4-8. A V-shaped watershed. (After Overton and Brakensiek [17]. Reprinted by permission of UNESCO and IAHS, © UNESCO–IAHS, 1973.)

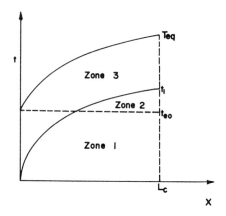

FIG. 4-9. Solution zones for V-shaped watershed in the characteristic plane of the channel. (Liggett and Woolhiser [16].)

of overland flow t_{eo}. Three possible solutions therefore exist and a complete analytical solution is not possible [16].

An approximate two-zone analytical solution was derived by Overton and Brakensiek [17]. The two zones are shown in Fig. 4-10. Unsteady overland flow is input to the channel in zone 1 and it is steady and equal to the rain rate between t_{eo} and T_{eq}, which is time to equilibrium of the entire watershed system. The kinematic equations are used to derive the channel flow hydrograph in a manner similar to that done for overland flow.

The results of the kinematic solution for the V-shaped watershed are shown in Table 4.1. The solutions in Table 4.1 can be placed in a convenient form by introducing the parameter μ_* as

$$\mu_* = t_{eo}/T_{eq} \tag{4-43}$$

and substituting it into the solution for the rising hydrograph, the two new equations are

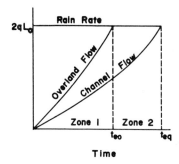

FIG. 4-10. Two-zone solution for rising hydrograph V-shaped watershed of Overton and Brakensiek [17]. Reprinted by permission of UNESCO and IAHS, © UNESCO–IAHS, 1973.

TABLE 4.1 Solution for V-shaped watershed[a]

Time period	Channel inflow	Rising hydrograph of channel Q_*
$0 \leq t \leq t_{eo}$	$2\,qL_0\,(t/t_{eo})^{5/3}$	$[(\tfrac{3}{8})(t/t_{eo})^{5/3}(t/t_{ec})]^{5/3}$
$t_{eo} \leq t \leq T_{eq}$	$2\,qL_0$	$[(t/t_{ec}) - (\tfrac{5}{8})(t_{eo}/t_{ec})]^{5/3}$
$t \geq T_{eq}$	$2\,qL_0$	1

[a] After Overton–Brakensiek [17]. Reprinted by permission of UNESCO and IAHS, © UNESCO–IAHS, 1973.

$$Q_* = \left\{ \frac{(t/T_{eq})^{8/3}}{[(5/3) + 1 - (5/3)\mu_*]\mu_*^{5/3}} \right\}^{5/3} \tag{4-44}$$

and

$$Q_* = \left\{ \frac{[(t/T_{eq}) - (5/8)]\mu_*}{1 - (5/8)\mu_*} \right\}^{5/3} \tag{4-45}$$

Equations (4-44) and (4-45) are plotted in Fig. 4-11 showing the effect of the parameter μ_*. When t_{eo} is very large relative to t_{ec} (which is often the case in the real world), the channel lagging effect is negligibly small. In this case Eq. (4-45) drops out and Eq. (4-44) becomes the solution.

$$Q_* = (t/T_{eq})^{40/9} \tag{4-46}$$

and the time to equilibrium of the watershed is closely approximated by t_{eo}.

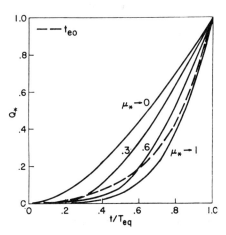

FIG. 4-11. Rising hydrograph studies of V-shaped watershed. (After Overton and Brakensiek [17].)

EXAMPLE 4.4   Estimate the rising and falling hydrograph for an impervious V-shaped watershed such as shown in Fig. 4-8 for a rain of 1 in./hr for 1 hr. The overland flow planes are asphalt ($n = 0.025$) and the dimensions are $L_0 = 300$ ft and $L_c = 300$ ft. Slope of the planes is 0.005. The ditch is 5 ft wide, made of concrete ($n = 0.015$) and sloped at 0.0025.

Using t_{eo} to approximate T_{eg} as Eq. (4-33):

$$t_{eo} = \frac{0.928}{1} \left[\frac{0.025(300)}{\sqrt{0.005}} \right]^{0.6} = 15.2 \text{ min}$$

Figure 4-11 can be used as quick approximation of the rising hydrograph (Fig. 4-12). Since the rain is for 1 hr, equilibrium will prevail for 44.8 min. The recession is calculated using Eq. (4-40) and Manning's equation.

4-9   Overland Flow on a Converging Surface

To account for the inaccuracies associated with the Wooding [15] V-shaped model, Veal [17] and Woolhiser [18] formulated and developed a model for concentrated flow which they called a converging surface. It is a watershed model consisting of a V-shaped section plus a portion of the surface of a cone as shown in Fig. 4-13.

The basic equations describing converging overland flow are

$$(\partial y/\partial t) + V(\partial y/\partial x) + y(\partial V/\partial x) = (q/w) + [Vy/(L_0 - x)] \qquad (4\text{-}47)$$

and

$$(\partial V/\partial t) + V(\partial V/\partial x) + g(\partial y/\partial x) = g(S_0 - S_f) - [q(V - u)/wy] \qquad (4\text{-}48)$$

where u is the velocity of the rain rate.

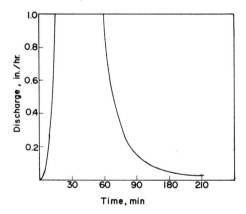

FIG. 4-12.

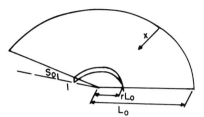

FIG. 4-13. A converging surface.
(After Woolhiser [19].)

Woolhiser [18] derived the kinematic equations for converging over-land flow. They are

$$dx/dt = amy^{m-1} \qquad (4\text{-}49)$$

and

$$dy/dt = (q/w) + [a(1-r)y^m/\{1-(1-r)x/L_0(1-r)\}] \qquad (4\text{-}50)$$

Therefore, the only difference between the converging surface equations and those for a plane is the convergence term in Eq. (4-50).

Since no general analytic solution has been obtained for these character-istic equations, numerical solutions have been necessary. The solutions for converging surfaces are shown in Fig. 4-14. As the convergence parameter r approaches unity, the solution approaches that of plane kinematic. As r approaches zero, the flow concentrates at a point at the end of the cone.

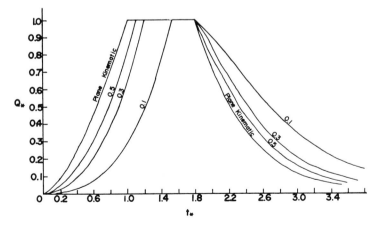

FIG. 4-14. Rising hydrograph and recession from equilibrium for the converging surface. Parameter is r. (After Woolhiser [19].)

4-10   Overland Flow on a Cascade of Planes

Kibler and Woolhiser [20] developed a kinematic solution for a cascade of planes such as shown in Fig. 4-15. The dimensionless characteristic equations for the kth plane are

$$dx_*/dt_* = \beta y_*^{m-1} \tag{4-51}$$

and

$$dy_*/dt_* = (q/wL_k)_* \tag{4-52}$$

where

$$\beta = m \sum_{i=1}^{k} l_i / \sum_{i=1}^{n} l_i = mL_k / \sum_{i=1}^{n} l_i \tag{4-53}$$

and n is total number of planes. Also,

$$q_* = Q_k/w \tag{4-54}$$

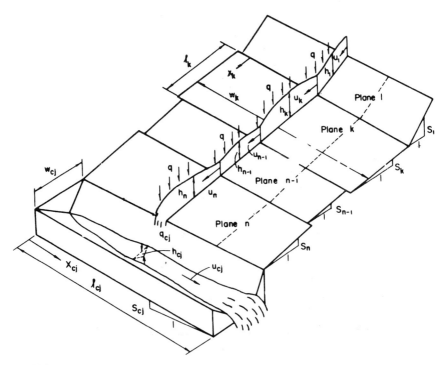

FIG. 4-15. Cascade of n planes discharging into jth channel section. (After Kibler and Woolhiser [20].)

where Q_k is the maximum discharge from downstream boundary of the kth plane.

Kibler and Woolhiser [20] solved Eqs. (4-51) and (4-52) for a hypothetical three-plane cascade of constant length and roughness but with varying combinations of slopes. The results of their computations are shown in Fig. 4-16 for a unit rain rate of 1 in./hr applied for 100 sec. The numerical solution was carried out using a single-step Lax–Wendroff scheme [21] which is an explicit method with a prescribed linear stability criteria. As shown in Fig. 4-16, the rising hydrographs have the same relative shapes as the slopes.

4-11 Kinematic Shock

For some combinations of slope, roughness, and width of a cascade of planes, a shock wave formation will occur. The surge causes successive waves to move with a greater speed so that earlier waves are overtaken and a shock wave, which is a coalescence of waves, is formed [20]. At the intersection of two characteristics, the depth-discharge relation is no longer valid, hence the path of shock must replace the rating.

The shock will occur at the intersection point when

$$\frac{dx}{dt}\bigg|_{\substack{\text{along upper} \\ \text{characteristic}}} > \frac{dx}{dt}\bigg|_{\substack{\text{along lower} \\ \text{characteristic}}} \tag{4-55}$$

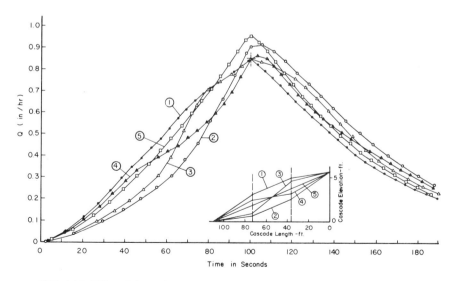

FIG. 4-16. Effect of slope-shape on hydrographs. (After Kibler and Woolhiser [20].)

This will occur when the shock parameter P_s is greater than unity [20].

$$(w_{k-1}/w_k)(a_{k-1}/a_k) = P_s > 1 \tag{4-56}$$

The local velocity of the shock wave will be

$$V_s = (\beta/m)(y_b^m - y_a^m)/(y_b - y_a) \tag{4-57}$$

where the subscripts a and b refer to the lower and upper characteristics A and B as shown in Fig. 4-17.

A number of shock wave solutions were traced out by Kibler and Woolhiser [20] and they found that the phenomenon was more mathematically induced than real. The single step Lax–Wendroff scheme was useful in smoothing out the shock which forces the rising hydrograph to abruptly rise and reach equilibrium.

EXAMPLE 4.5   For the following two-plane cascade, determine if a kinematic shock will be induced on the lower plane. The width of the planes are the same: $n_a = 0.08$, $L_a = 200$ ft, $S_a = 0.001$, $n_b = 0.025$, $L_b = 200$ ft, $S_b = 0.01$.

From Eq. (4-8),

$$a_1 = \frac{1.49}{0.08}\sqrt{0.01} = 1.86 \quad \text{and} \quad a_2 = \frac{1.49}{0.025}\sqrt{0.001} = 1.88$$

and from Eq. (4-56),

$$P_s = \frac{1.86}{1.88} = 0.989$$

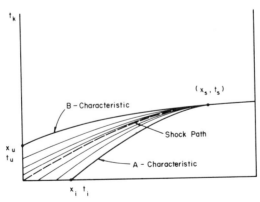

FIG. 4-17. Shock-wave path given by locus of intersecting characteristics. (After Kibler and Woolhiser [20].)

Hence, even though the upper plane is steeper than the lower plane, no shock will be induced because the roughness of the upper plane slows the flow down such that the B and A characteristics do not intersect.

4-12 Kinematic Streamflow

Lighthill and Whitham [22] have shown that for $\mathbf{F} = 2$, kinematic waves predominate over dynamic waves, and generally a kinematic condition occurs in supercritical flow. However, Gburek and Overton [23] have shown that subcritical kinematic streamflow occurred during a three-month storm period in a 7 mile stretch of a 162 mile^2 watershed tributary to the Susquehanna in Pennsylvania. Froude numbers were never higher than 0.34. The hydraulic reasons for this phenomenon were not determined, but as Lighthill and Whitham [22] have pointed out, kinematic flow is occurring when a single-valued rating between area and discharge exists in a stream.

Using the single-valued rating for a ractangular channel

$$Q = A_0 y^b \qquad (4\text{-}58)$$

it can be shown that

$$\mathbf{F} = \frac{A_0 y^{b-3/2}}{w \sqrt{g}} \qquad (4\text{-}59)$$

$$\partial V/\partial x = (A_0/w)(b-1)y^{b-2}(\partial y/\partial x) \qquad (4\text{-}60)$$

$$\partial V/\partial t = -b(\partial y/\partial x) + (q/Vw)(b-1)\mathbf{F}^2 \qquad (4\text{-}61)$$

and

$$(V/g)(\partial V/\partial x) = (b-1)\mathbf{F}^2(\partial y/\partial x) \qquad (4\text{-}62)$$

Neglecting the local inflow term and utilizing Eqs. (4-58)–(4-62), the momentum equation (3-17) becomes

$$S_0 - S_f = \{1 - (b-1)^2 \mathbf{F}^2\}(\partial y/\partial x) \qquad (4\text{-}63)$$

which leads to

$$Q = Q_n[1 - (\partial y/\partial x/S_0)\{1 - (b-1)^2 \mathbf{F}^2\}]^{1/2} \qquad (4\text{-}64)$$

Hence, Eq. (4-64) can be used to quickly evaluate whether or not the dynamic terms in the momentum equation (4-64) are appreciably large.

EXAMPLE 4-6 In the study of Gburek and Overton [23] of the Mahantango Creek, the maximum value of $\partial y/\partial x$ relative to bed slope S_0 was 0.14

and was associated with a maximum Froude number of 0.34. This was observed at the upstream gauge at the peak discharge of 543 cfs. The exponent b on the rating curve established at this station was 2.74.

Evaluate the value of the momentum terms in Eq. (4-64)

$$Q - Q_n\{1 - 0.14\{1 - (2.74 - 1)^2(0.34)^2\}\}^{1/2}$$

and

$$Q = Q_n(1 - 0.091)^{1/2} = 0.953Q_n$$

It follows that the flow is very nearly uniform and the kinematic approximation should be an accurate representation of the inbank streamflow regime of this creek. Solution for future streamflow computations can be obtained by simultaneous solution of Eqs. (3-6) and (4-58).

Another method of evaluating Eq. (4-64) is to examine the ratio of kinematic to dynamic wave speed and this can be shown to be

$$\frac{(1/w)(dQ/dy)}{dx/dt} = \frac{bV}{V + \sqrt{gy}} = \frac{b\mathbf{F}}{\mathbf{F} + 1} \qquad (4\text{-}65)$$

Although kinematic wave speed reached a maximum of only 70% of the dynamic wave speed in the study of Gburek and Overton [23], nevertheless the flow was predominantly kinematic.

EXAMPLE 4-7   For the flow condition in Example 4.6, estimate the kinematic wave speed relative to the dynamic wave speed.

Using Eq. (4-65),

$$\frac{\text{kinematic WS}}{\text{dynamic WS}} = \frac{2.74(0.34)}{0.34 + 1} = 0.70$$

Hence, kinematic wave speed reached a maximum of 70% of dynamic wave speed over the period of record.

The characteristic equations for kinematic streamflow where lateral inflow is not a significant part of the streamflow volume are

$$dx/dt = V + \sqrt{gy} \qquad (4\text{-}66)$$

and

$$(1/g)(d\sqrt{gy}/dt) = \tfrac{1}{2}\{1 - (b - 1)^2\mathbf{F}^2\}\,\partial y/\partial x \qquad (4\text{-}67)$$

If single-valued rating curves are known from field observation, then Eqs.

(4-66) and (4-67) may be integrated without reference to a Manning n-value. And, if the acceleration terms are negligibly small, Eq. (4-67) simplifies to

$$(1/g)(d\sqrt{gy}/dt) = \tfrac{1}{2}\,\partial y/\partial x \qquad (4\text{-}68)$$

EXAMPLE 4.8   For the Mahantango Creek, estimate the depth of flow 1.1 miles downstream from the uppermost streamflow station using Eq. (4-68) and the implicit grid system in Fig. 3-6 where $y_1 = 3.47$ ft, $y_2 = 4.20$ ft, $y_3 = 3.53$ ft, and $\Delta t = 2$ hr.

The finite difference form of Eq. (4-68) becomes

$$y_4 - (2\,\Delta x/g\,\Delta t)\sqrt{gy_4} - (2\,\Delta x/\sqrt{g}\,\Delta t)\{\sqrt{y_2} - \sqrt{y_3} - \sqrt{y_1}\}$$
$$+ y_3 - y_2 - y_1 = 0$$

and for the problem at hand is

$$y_4 + 0.284\sqrt{y_4} - 3.66 = 0$$

The solution for y_4 is

$$y_4 = 4.25 \text{ ft}$$

There is evidence, therefore, both theoretical and experimental that streamflow does occur in a kinematic state in natural streams even where local or lateral inflow is not a substantial contributor. Most conditions of stormwater modeling will be where lateral inflow is the main contributor, as in the V-shaped watershed example. For conditions where lateral inflow is the main contributor, the V-shaped watershed model may be used. Where the volume associated with the upstream hydrograph is large relative to the lateral inflow volume, Eqs. (4-66) and (4-67) or (4-68) are applicable if the rating function is known.

4-13   Free Surface Storm Sewer Flow

The kinematic approximation is applicable to free surface storm sewer flow, i.e., in contrast to pressure flow in pipes. The basic concept used in design of storm sewers is that the flow is controlled by gravity since pressurized sewers often cause problems such as pollution through leaky joints and foundation washout.

For free surface pipe flow, the equation of motion for the kinematic approximation can be placed in the form of

$$Q = aA^m(y) \qquad\qquad (4\text{-}69)$$

where y is the depth of flow in the pipe and $A(y)$ is the associated area of flow. Solution of the stormwater hydrograph for a catchment-sewer system proceeds as in the V-shaped watershed. The concentration effect as shown for the V-shaped watershed is also applicable to a sewered catchment but the effect will be more pronounced due to the very low roughness of concrete pipe ($n = 0.01 - 0.015$). Hence, a storm sewer has the effect of getting water away from a catchment faster than would be the case for an open ditch or channel which was not lined with concrete.

There are other methods used in the design of storm sewers. Yen and Sevuk [24] attempted a comparison of several methods, including the kinematic wave model, used in storm sewer design. Most of the methods are considerably less direct and more complex than the kinematic wave model and would require more computations. They found that using the complete equations of motion resulted in a more accurate storm sewer design than would be obtained by the kinematic wave model. However, it was not made entirely clear exactly how much accuracy was sacrificed relative to the cost and time savings associated with the use of the kinematic approximation.

Problems

4-1. A 500-ft-wide, 1000-ft-long river reach is flowing such that it is approximately wide rectangular. Assume flow is nearly kinematic and that the depth of flow is 5 ft. If the Manning n-value is 0.05 and bed slope is 0.001 estimate
  (a) the speed of the wave relative to an observer standing on the bank, and
  (b) the time of arrival of the wave traveling from the upper to the lower end of the reach.
Can you draw any conclusions from this exercise?

4-2. True or false
  (a) Critical depth is a function of channel slope.
  (b) Momentum is a vector.
  (c) Energy losses are attributable to skin drag.
  (d) There is no backwater in supercritical flow.
  (e) Internal forces are not considered in Newton's three laws of motion.

4-3. Name two types of phenomena which can account for momentum losses.

4-4. When can we safely assume the pressure distribution coefficient to be unity?

4-5. What is the speed of a wave in a stream relative to an observer standing on the bank equal to?

4-6. What is kinematic flow?

4-7. Why isn't the Euler number an important consideration in open channel flow?

4-8. Derive a general mathematical expression for the hydrograph at the end of a uniform asphalt plane 400 ft long sloped at 0.0025. Assume $n = 0.025$. The time distribution of rain intensity i is sinusoidal but uniform.

$$i(t) = i_0 \sin(\pi t/T) \qquad 0 \le t \le T$$

where $i_0 = 1$ in./hr and $T = 30$ min. Assume that the Manning-kinematic approximation accurately approximates the equations of motion for all flow. Derive the expression between $0 \le t \le T$ only. Solve for the peak flow from above.

4-9. Derive a criteria for when Manning-kinematic overland flow is subcritical on a uniform plane under steady uniform rain.

4-10. Given the sinusoidal overland flow plane feeding into the wide-rectangular channel shown in Fig. 4.18.

(a) Derive an expression for the rising hydrograph at the end of the channel for a long steady, uniform unit input. If an explicit relation cannot be derived, then indicate how it can be derived. Use the Manning-kinematic approximation.

$$L(x) = L_0 \sin(\pi x/L_c)$$

(b) Derive an expression for:
   (1) equilibrium depth profile in the channel.
   (2) time of concentration of the entire catchment-channel system.

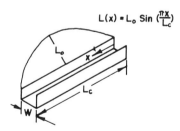

$$L(x) = L_0 \, Sin \left(\frac{\pi x}{L_c}\right)$$

FIG. 4-18.

4-11. Determine the range of steady-uniform rainfall intensities for which the Manning-kinematic solution is a highly accurate approximation to the rising hydrograph on a uniform concrete plane 50 ft long sloped at 0.5% (let $n = 0.021$).

4-12. (a) In the two parking lots near Knoxville shown in Figs. 4-19 and 4.20, estimate the peak discharge at points A generated by a rain storm with a return period equal to 5 yr. Use the Manning-kinematic solution. The parking lots are asphalt ($n = 0.025$) and both are sloped at 0.0025.

(b) Would the answer for Fig. 4.19 have been the same if the 400 and 200 ft planes had been interchanged? Prove your answer.

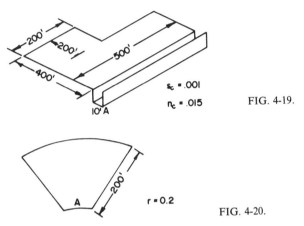

FIG. 4-19.

FIG. 4-20.

4-13. What is the theoretical justification for the use of the Manning and laminar resistance laws in the kinematic wave approximation in view of the fact that it was derived for Chezy resistance law?

4-14. Generally, what is the effect of rain intensity and Reynolds number on the laminar-turbulent transition on overland flow?

4-15. What is kinematic overland flow?

4-16. Describe the nature of the kinematic shock using as few equations as necessary.

4-17. What are the advantages of Lax–Wendroff scheme over others tested by Kibler and Woolhiser?

4-18. Write a computer program to calculate flow in a storm sewer using the kinematic approximation where inflow is from the overlying parking lot and enters the sewer through stormwater inlets and lateral storm sewers.

4-19. Considering one-dimensional flow only, does a characteristic have a physical interpretation? If so, what is it? What is the mathematical significance of a characteristic curve in the x, t plane?

4-20. An orthogonal network of characteristic curves can be constructed (graphically or by computation) in the distance–time plane for certain kinds of fluid motion. Physically, how is this network related to the actual flow, i.e., what do the different sets of curves represent?

4-21. What significance, if any, would you attach to the intersection of two characteristics curves which have the same general slope in the x, t plane?

4-22. As a computational method of solving an unsteady flow problem, what is the main advantage of the method of characteristics over other solution techniques for mathematical flow models. What is the main disadvantage of the method from a practical point of view?

4-23. In what ways are streamflow routing in channels different from the modeling of overland flow? Cite both physical differences and the resulting general differences in mathematical models and solution procedures.

4-24. Is kinematic flow routing applicable to rapidly varied flow? Why or why not?

4-25. Calculate the stormwater hydrograph at the end of a 300 ft plane for the following rainfall intensity time distribution:

Time (min):	0–5	5–10	10–15
Rain intensity (in./hr):	2.0	0	1.0

The concrete plane is uniformly sloped at 0.005.

4-26. Derive Eqs. (4-19) and (4-22).

4-27. Derive Eqs. (4-28)–(4-30).

4-28. Derive Eqs. (4-40) and (4-41).

4-29. Derive the equations for the V-shaped watershed in Table 4.1.

4-30. Derive Eqs. (4-47) and (4-48).

4-31. Derive Eqs. (4-51), (4-52), and (4-53).

4-32. Derive the expression for local shock wave velocity, Eq. (4-57).

4-33. Derive the kinematic streamflow model, Eq. (4-64).

4-34. Derive Eq. (4-68).

References

1. Henderson, F. M., "Open Channel Flow," pp. 288–293. Macmillan, New York, 1966.
2. Woolhiser, D. A., and Liggett, J. A., Unsteady one-dimensional flow over a plane—the rising hydrograph, *Water Resources Res.* **3**, No. 3, 753–771 (1967).
3. Henderson, F. M., and Wooding, R. A., Overland flow and groundwater from a steady rainfall of finite duration, *J. Geophys. Res.* **69**, No. 8, 1531–1540 (April 1964).
4. Liggett, J. A., and Woolhiser, D. A., Difference solutions of the shallow water equations, *Proc. ASCE J. Mech. Div.,* 39–71 (April 1967).
5. Corps of Engineers, U.S. Army, Data report, airfield drainage investigations, Los Angeles District, Office of the Chief of Engr., Airfields Branch Eng. Div., Military Construction, Oct. 1954.
6. Overton, D. E., A variable response overland flow model, Ph.D. Dissertation, Univ. of Maryland, Dept. of Civil Eng., 1971.

7. Schreiber, D. L., and Bender, D. L., Obtaining overland flow resistance by optimization, *J. Hydr. Div., Amer. Soc. Civil Engr.* **98**(HY3), 429–446 (1972).

8. Shen, Hsieh, W., and Ruh-Ming Li, Rainfall effect on sheet flow over smooth surface, *J. Hydr. Div., Amer. Soc. Civil Engr.* **99**(HY5), 771–792 (1973).

9. Yu, Y. S., and McKnown, J. S., Runoff from impervious surfaces, Contract Rep. No. 2–66, US Army Engr. Waterways Expt. Station, Vicksburg, Mississippi, 1963.

10. Morgali, J. R., Laminar and turbulent overland flow hydrographs, *J. Hydr. Div., Amer. Soc. Civil Engr.* **96**(HY2), 441–460 (Feb. 1970).

11. Overton, D. E., Kinematic flow on long impermeable planes, *Water Resources Bull.* **8**, No. 6, 1198–1204 (1972).

12. Overton, D. E., Route or convolute? *Water Resources Res.* **6**, No. 1, 43–52 (1970).

13. Horton, R. E., The interpretation and application of runoff plot experiments with reference to soil erosion problems, *Soil Sci. Soc. Amer.* **3**, 340–349 (1938).

14. Izzard, C. F., Hydraulics of runoff from developed surfaces, *Proc. 26th Ann. Meeting Highway Res. Board, Washington D.C.*, 129–150 (1946).

15. Wooding, R. A., A hydraulic model for the catchment stream problem, *J. Hydrol.* (a) Kinematic-wave theory **3**, 254–257 (1965); (b) Numerical solutions **3**, 268–282 (1965); (c) Comparison with runoff observations **4**, 21–37 (1966).

16. Liggett, J. A., and Woolhiser, D. A., The use of the shallow water equations in runoff computation, *Proc. Third Annual Amer. Water Resources Conf., San Francisco*, 117–126 (1967).

17. Overton, D. E., and Brakensiek, D. L., A kinematic model of surface runoff response, *Proc. Symp. Results Res. Representative Exptl. Basins, Wellington, New Zealand, Dec. 1970* **1**. IASH–UNESCO, Paris, 1973.

18. Veal, D. G., A computer solution of converging overland flow, M.S. Thesis, Cornell Univ., New York, 1966.

19. Woolhiser, D. A., Overland flow on a converging surface, *Trans. Amer. Soc. Agr. Engr.* **12**, No. 4, 460–462 (1969).

20. Kibler, D. F., and Woolhiser, D. A., The kinematic cascade as a hydrologic model, Hydrology Papers, Colorado State Univ., No. 39, March 1970.

21. Houghton, D. D., and Kasahara, A., Nonlinear shallow fluid flow over an isolated ridge, *Commun. Pure Appl. Math.* **XXI** (1968).

22. Lighthill, M. J., and Whitham, G. B., Kinematic waves 1, *Proc. Roy. Soc., London A* **229**, 281–316 (1955).

23. Gburek, W. J., and Overton, D. E., Subcritical kinematic flow in a stable stream, *J. Hydr. Div., Amer. Soc. Civil Engr.*, **99**(HY9) 1433–1447 (1973).

24. Yen, B. C., and Sevuk, A. S., Design of storm sewer networks, *J. Environ. Engr., Amer. Soc. Civil Engr.* **101**, No. EE4, 535–553 (Aug. 1975).

Chapter 5 | Estimation of Time of Concentration Using Kinematic Wave Theory

5-1 Introduction

The primary surface water system response parameter which emerged from the kinematic approximation was time to equilibrium t_e. This parameter appears to be quite similar to the much used "time of concentration" in the so-called rational method of peak runoff design.

Time of concentration is allegedly the time that a particle of water in the most remote part of the watershed takes to reach the watershed outlet. Time to equilibrium is equal to

$$t_e = L/V_0 \tag{5-1}$$

and if L is the length of the plane, the system reaches equilibrium if a particle of water traveled that distance at a speed equal to V_0. But as we saw from kinematic wave theory in Chapter 4, the velocity of the wave is increasing as it travels the plane from zero to a maximum of V_0 at the outlet. Hence, the average speed of the kinematic wave is

$$\frac{1}{t_e} \int_0^{t_e} \left(\frac{dx}{dt} \right) dt = \frac{am}{t_e} \int_0^{t_e} (qt)^{m-1} \, dt \tag{5-2}$$

and the average is

$$\overline{dx/dt} = V_0/m \tag{5-3}$$

For turbulent flow the average wave speed is $0.6V_0$, hence, the conceptual time of concentration relative to time to equilibrium is

$$t_c/t_e = 1.67 \qquad (5\text{-}4)$$

It is concluded that if peak runoff designs are based upon the classical concept of time of concentration, the design will not be for the maximum condition since it takes a time t_e to reach equilibrium. Therefore, it is proposed to use t_e in this book as being synonymous with time of concentration. In that case, a complete hydraulic solution is at hand for deriving this runoff parameter.

In this chapter, t_c will be derived for several catchment and small watershed geometries, and guidelines will be shown for deriving the parameter for any complex basin shape. Most of the concepts in this chapter have been previously reported by Overton [2].

5-2 Derivation of Time of Concentration

As shown in Chapter 4, i.e., Fig. 4-2 and Eq. (4-25), lagtime is defined as

$$t_L = S_\infty/q/w \qquad (5\text{-}5)$$

Equation (5-5) is applicable to any basin shape so long as the supply rate q is steady and uniform. Further, it was shown by Eq. (4-29) that lagtime is a constant factor of time of concentration. For turbulent flow,

$$t_c = 1.6t_L \qquad (5\text{-}6)$$

Therefore, t_c can be derived for any basin by evaluating the equilibrium storage S_∞ and dividing it by the supply rate q/w.

5-3 Time of Concentration for a Plane

At equilibrium condition, the transient term in Eq. (3-6) drops out, and using the Manning formula for the normal flow relation, the depth profile can be found by solving Eqs. (4-26) and (3-23):

$$y(x) = \left[\frac{qn}{w1.49\sqrt{S_0}} x \right]^{0.6} \qquad (4\text{-}34)$$

Then the average depth at equilibrium S_∞ is

$$S_\infty = \frac{1}{L} \int_0^L y(x)\, dx = \tfrac{5}{8} \left[\frac{qnL}{w1.49\sqrt{S_0}} \right]^{0.6} \qquad (5\text{-}7)$$

and the solution for t_c is

$$t_c = [0.928/(i)^{0.4}](nL/\sqrt{S_0})^{0.6} \tag{5-8}$$

where i is in inches per hour and t_c is in minutes; Eq. (5-8) can be used to physically derive time of concentration.

5-4   Time of Concentration of a Cascade of Planes

Consider the cascade of planes shown in Fig. 5-1. The equilibrium storage for the cascade is determined by integrating the depth profile over each plane. The general relation for p-planes is

$$S_\infty = \sum_{j=1}^{p} \left(\frac{n(q/w)}{1.49 \sqrt{S_0}} \right)_j^{0.6} \frac{(L_j^{1.6} - L_{j-1}^{1.6})}{1.6L_p} \tag{5-9}$$

The time of concentration is then found by multiplying Eq. (5-9) by $1.6/(q/w)$.

Remembering from Chapters 3 and 4 that in zone A the solution is not a function of the upstream or downstream condition, it follows that a completely general expression for S_∞ for any spatial distribution of supply rate can be found by calculating equilibrium storage from Eq. (5-9) with the supply rate on each plane, and then determine time of concentration as

$$t_c = \frac{1.6S_\infty}{\sum_{j=1}^{p} (q_j(L_j - L_{j-1})/w)/L_p} \tag{5-10}$$

EXAMPLE 5.1   Estimate the time of concentration for a cascade of three planes being subjected to a unit rain excess intensity. The following data are available: From the top to the bottom of the hill the lengths are 150,200, and 100 ft; the slopes are 0.001, 0.030, and 0.002, respectively. The hillslope is in closely cut rye grass.

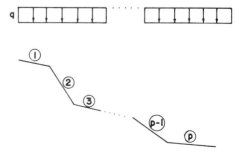

FIG. 5-1. A cascade of planes.

Using an n-value of 0.085 and Eq. (5-9),

$$S_\infty = \left[\frac{0.085(q/w)}{1.49}\right]^{0.6}\left[\frac{(150)^{1.6}}{(0.001)^{0.3}} + \frac{(350)^{1.6} - (150)^{1.6}}{(0.030)^{0.3}}\right.$$

$$\left. + \frac{(450)^{1.6} - (350)^{1.6}}{(0.002)^{0.3}}\right]\frac{(0.625)}{450}$$

The rain rate (cfs/ft^2) is

$$\frac{q}{w} = \frac{1}{43908} = 2.28 \times 10^{-5} \text{ ft/sec}$$

and

$$S_\infty = 0.035 \text{ ft}$$

Then, $t_c = (1.6)(0.035)/2.28 \times 10^{-5} = 2456$ sec $= 40.9$ min.

5-5 Time of Concentration of a V-Shaped Watershed

Referring to Fig. 4-10, the time of concentration for the V-shaped watershed is

$$t_c = t_{eo} + t_{ec} \tag{5-11}$$

In terms of physical and hydrologic characteristics, Eq. (5-11) becomes

$$t_c = [0.928/(i)^{0.4}][(nL/\sqrt{S_0})_0^{0.6} + (w/2L_0)^{0.4}(nL/\sqrt{S_0})_c^{0.6}] \tag{5-12}$$

The convergence factor $(w/2L_0)^{0.4}$ renders the time of concentration of the channel very small relative to that of the plane.

EXAMPLE 5.2 For the V-shaped watershed in Fig. 4-8, calculate the convergence factor and the time of concentration for a unit rain excess rate. Is t_c for the channel negligibly small? The following data are available:

$$L_0 = 400 \text{ ft}, \qquad n_0 = 0.03, \qquad L_c = 1000 \text{ ft}$$

$$n_c = 0.045, \qquad S_0 = 0.05, \qquad S_c = 0.001, \qquad w = 20 \text{ ft}.$$

The convergence factor is

$$[20/2(400)]^{0.4} = 0.23$$

and

$$t_c = \frac{0.928}{1} \left[\left(\frac{(0.03)(400)}{\sqrt{0.05}} \right)^{0.6} + 0.23 \left(\frac{(0.045)(1000)}{\sqrt{0.001}} \right)^{0.6} \right]$$

$$= 0.928(10.9 + 17.9) = 26.8 \text{ min}$$

t_c for the channel is not negligibly small. Actually it is larger, being 16.6 min as compared to 10.1 min for overland flow.

5-6 Time of Concentration of a Converging Surface

Reference is made to Fig. 4-13. At the equilibrium condition, Eq. (4-47) reduces to

$$dQ/dx = q + [Q/(L_0 - x)] \tag{5-13}$$

and the solution of Eq. (5-13) for discharge is

$$Q = q(L_0 - \tfrac{1}{2}x)x/(L_0 - x) \tag{5-14}$$

By eliminating discharge from Eqs. (3-23) and (5-14), the solution for the depth profile is

$$y(x) = \left[\left(\frac{nq/w}{1.49 \sqrt{S_0}} \right) \frac{(L_0 - \tfrac{1}{2}x)x}{L_0 - x} \right]^{0.6} \tag{5-15}$$

For convenience, Eq. (5-16) was placed in dimensionless terms

$$y_* = \frac{y}{\{nL_0(q/w)/1.49 \sqrt{S_0}\}^{0.6}} \tag{5-16}$$

and

$$x_* = x/L_0 \tag{5-17}$$

Then, Eq. (5-15) becomes

$$y_* = \frac{(1 - 0.5x_*)x_*}{1 - x_*} \tag{5-18}$$

The average dimensionless depth of flow is defined as

$$\bar{y}_* = \frac{1}{1 - r} \int_0^{1-r} y_* \, dx_* \tag{5-19}$$

An exact solution of Eq. (5-19) could not be found, and the integral was evaluated numerically. For values of r less than 0.05 approaching zero, the depth at the outlet of the converging surface becomes extremely large. The dimensionless mean depth of flow has been plotted versus $1 - r$ in Fig. 5-2 showing the slight curvature of the line. It was found that the curve could be very accurately represented by the line of equal values. Only a relative standard error of 6.4% resulted with an associated simple correlation coefficient of 0.997.

Equation (5-19) is then approximated as

$$\bar{y}_* \approx (1 - r) \qquad 0.05 \leq r \leq 1 \tag{5-20}$$

and the mean depth over the converging surface at equilibrium is approximately

$$S_\infty = (1 - r)[nL_0(q/w)/1.49 \sqrt{S_0}]^{0.6} \tag{5-21}$$

The time of concentration becomes

$$t_c = [0.928(1 - r)/i^{0.4}](nL_0/\sqrt{S_0})^{0.6} \tag{5-22}$$

In terms of the actual length of the flow path $L = (1 - r)L_0$, Eq. (5-22) can be written as

$$t_c = 0.928[(1 - r)/i]^{0.4}(nL/\sqrt{S_0})^{0.6} \tag{5-23}$$

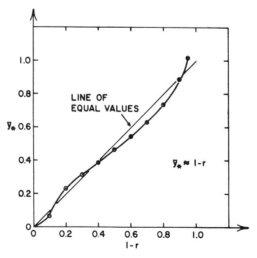

FIG. 5-2. Solution of dimensionless depth of flow on a converging surface as a function of $1 - r$. (After Overton [2].)

EXAMPLE 5.3   An asphalt parking lot can be approximated as the converging surface in Fig. 4-13. Estimate the time of concentration for a unit rain excess intensity if the following data are available:

$$n = 0.028, \qquad S_0 = 0.005, \qquad L_0 = 300 \text{ ft}, \qquad r = 0.20$$

Using Eq. (5-23), we find

$$t_c = 0.928 \left(\frac{1 - 0.20}{1} \right)^{0.4} \left(\frac{(0.028)(300)}{\sqrt{0.005}} \right)^{0.6} = 14.9 \text{ min}$$

The equilibrium storage is

$$S_\infty = \frac{1 \, (\text{in/hr}) \times (14.9/60) \, \text{hr}}{1.6} = 0.25 \text{ in}$$

5-7   A Concept of Lag Modulus

The solutions for lagtime that have been derived for the various geometries are in terms of three basic variables—geometry, roughness, and input. Geometry and roughness can be lumped into a single physical variable μ which is designated as a lag modulus. For the plane surface,

$$\mu_P = 0.928(nL/\sqrt{S_0})^{0.6} \tag{5-24}$$

μ is designated as a lag modulus because it is a real positive number that expresses the lag or surface runoff response in terms of the physics of the flow systems for a unit input $i = 1$. As long as the physics of the system are time-invariant, the lag or surface water response is only a function of input rate, and a different unit hydrograph would result for each rain intensity. For a catchment with a fixed lag modulus, the time of concentration could be generalized as

$$t_C = f(\text{input}) \tag{5-25}$$

The variability of response with input rate for overland and watershed runoff has been reported by Amorocho and Orlob [3], Overton [4], and Overton and Brakensiek [5].

By comparison of the lag moduli for a plane surface and for a converging surface,

$$\frac{\mu_C}{\mu_P} = \frac{0.928(1 - r)^{0.4}(nL/\sqrt{S_0})_C^{0.6}}{0.58(nL/\sqrt{S_0})_P} \tag{5-26}$$

it is clearly shown that the two moduli are related by a factor if the physical characteristics n, L, and S_0 are the same.

$$\frac{\mu_C}{\mu_P} = 1.6(1 - r)^{0.4} \qquad (5\text{-}27)$$

The ratio of the modulus of the converging surface to the modulus of a plane surface is designated as a convergence factor σ.

$$\sigma = \frac{\mu_C}{\mu_P} \qquad (5\text{-}28)$$

Again, in an attempt to generalize and extrapolate these simple solutions to more realistic catchment shapes for urban areas (streets, parking lots, etc.) and rural watersheds, time of concentration can be written as

$$t_C = \sigma \mu_P i^{-0.4} \qquad (5\text{-}29)$$

In the following section, a method will be derived for operating on observed input and output data from a complex catchment in an attempt to experimentally derive convergence factors.

5-8 Estimation of Time of Concentration on Complex Catchments Where Stormwater Data Are Available

Recognizing that there will be errors created by averaging the physical characteristics of catchments that will be lumped into the experimental determinations of the conveyance factor, we will, however, proceed to develop a numerical procedure for evaluating σ for complex storms generated on physically complex catchments. Consider the hypothetical runoff hydrograph shown in Fig. 5-3 and the time distribution of input. Assume that this is a catchment with some degree of nonuniform convergence. If this were strictly a linear system, the lagtime for each unit hydrograph associated with each block of input would be identical. This lagtime would also equal the overall lagtime of the storm. However, by the model here, each unit response has a different lagtime. It follows that the time of concentration for the entire storm should be a weighted average of the lagtime for each input.

$$\bar{t}_L = \frac{i_1 t_{L_1} + \cdots + i_K t_{LK}}{i_1 + \cdots + i_K} \qquad (5\text{-}30)$$

Inserting Eq. (5-29) into (5-30) and factoring out convergence factor and lag modulus,

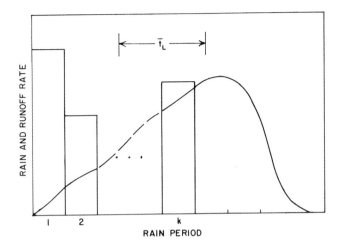

FIG. 5-3. Hypothetical complex runoff event for derivation of convergence factor. (After Overton [2].)

$$\bar{t}_c = \sigma \mu_P \frac{i_1^{0.6} + \cdots + i_K^{0.6}}{i_1 + \cdots + i_K} \tag{5-31}$$

Then, solving for the convergence factor times lag modulus, we obtain

$$\sigma \mu_P = \frac{\bar{t}_c \sum_1^K i}{\sum_1^K i^{0.6}} \tag{5-32}$$

remembering that μ_P is some averaged lag modulus of a plane.

EXAMPLE 5.4   Estimate the convergence factor and the lag modulus for a parking lot which resembles the converging surface in Fig. 4-11 from the given storm data.

Weighted average storm lagtime determined from the hyetograph and storm hydrograph was 20.7 min and the rainfall hyetograph is as shown in Table 5-1. Using Eq. (5-33), we obtain

$$\sigma \mu_p = \frac{(7.7)(1.6)(20.7)}{5.71} = 44.7 \text{ min}$$

Hence, this experimentally determined time of concentration may be used in stormwater predictions associated with design rainfall events on this watershed for the same land use pattern.

TABLE 5-1

Time (min)	Rain intensity (in./hr)	$i^{0.6}$
0–5	2.7	1.81
5–10	1.2	1.12
10–15	0.7	0.81
15–20	0.7	0.81
15–20	3.1	1.97
	$\sum i = 7.7$	$\sum i^{0.6} = 5.71$

5-9 Summary

The lagtime presented by Overton [2] can be extrapolated to estimate lagtime for natural and/or engineered catchments as long as one can arrive at a Manning n-value(s). However, one must realize that the catchment parameters must be averaged in some manner, and the averaging induces errors into the lagtime estimation.

It was shown that for a given lag modulus and input, lagtime for a converging surface is a factor of lagtime for a plane. The lagtime and time of concentration in general for any catchment can then be written as a convergence factor times lagtime for a plane.

For further surface water investigations, a numerical procedure was developed whereby the convergence factor can be calculated by operating on the input and output of the catchment.

Problems

5-1. Estimate the maximum depth of flow at the end of a 500 ft asphalt runway near Knoxville generated by a rain storm with a 10 yr return period ($n = 0.025$; $S = 0.005$).

5-2. Given the impervious two-plane cascade shown in Fig. 5.4, will a kinematic shock result from a uniform rain intensity of (a) 1 in./hr? (b) 2 in./hr?

FIG. 5-4.

5-3. Contrast time to equilibrium with the classical definition of time of concentration both conceptually and mathematically.

5-4. Derive Eqs. (5-9) and (5-10).

5-5. Derive Eq. (5-15).

5-6. (a) Determine a general relationship for equilibrium storage, S_∞, and lagtime t_L for a converging surface as a function of r.

 (b) Calculate S_∞ and t_L using the relation in (a) above for

$$r = 0.10, \quad n_c = 0.02, \quad L_0 = 500 \text{ ft}, \quad S_0 = 0.01, \quad i = 1 \text{ in./hr}$$

5-7. Determine t_c for the two-plane cascade (Fig. 5-5) for two rain patterns

 (1) $i(a) = 1$ in./hr, $i(b) = 0.5$ in./hr

 (2) $i(a) = 0.5$ in./hr, $i(b) = 1$ in./hr

Sketch the results.

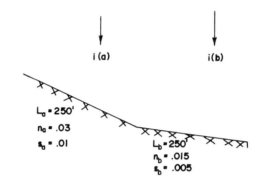

FIG. 5-5.

5-8. Estimate time of concentration for a V-shaped watershed for a unit rain excess intensity using the following physical data:

$$L_0 = 300 \text{ ft}, \quad L_c = 300 \text{ ft}, \quad n_0 = 0.025, \quad n_c = 0.045$$
$$w = 10 \text{ ft}, \quad S_0 = 0.023, \quad S_c = 0.005.$$

References

1. Kuichling, E., The relation between the rainfall and the discharge of sewers in populous districts, *Trans. Amer. Soc. Civil Engr.* **20**, 1–56 (1889).
2. Overton, D. E., Estimation of surface water lagtime from the kinematic wave equations, *Water Resources Bull.* **7**, No. 3, 428–440 (June 1971).
3. Amorocho, J., and Orlob, G. T., Nonlinear analysis of hydrologic systems, Water Res. Center Contr. No. 40, Univ. California, Berkley, 1961.
4. Overton, D. E., Analytical simulation of watershed hydrographs from rainfall. *Proc. Intern. Hydrol. Symp., Ft. Collins, Colorado* (1967).
5. Overton, D. E., and Brakensiek, D. L., A kinematic model of surface runoff response, *Symp. Results Res. Representative Exptl. Basins, Wellington, New Zealand, Dec. 1970* **1**. IASH–UNESCO, Paris, 1973.

Chapter 6 | Examples of Deterministic Stormwater Modeling

6-1 Introduction

This chapter contains examples of deterministic modeling of rural and urban stormwater. These models were built using a linkage of the component models described in the previous chapters. Deterministic modeling of rural stormwater preceded that of urban stormwater by nearly a decade and received a substantial impetus from the advent of the second generation high speed digital computer around 1961.

Urban stormwater modeling received a large stimulus in the late 1960s with the creation of the Urban Water Resources Program of the American Society of Civil Engineers [1]. Much of the pioneering work in deterministic modeling of urban stormwater systems was done by the firms of Water Resources Engineers, Inc., Metcalf and Eddy, Inc., the University of Florida, and the Massachusetts Institute of Technology. All of these models rely heavily upon the kinematic wave approximation to the surface runoff process.

6-2 A Model of Rural Stormwater

Smith and Woolhiser [2] have reported a deterministic stormwater model composed of a linkage between Richard's equation (2-17) and the kinematic cascade model developed by Kibler and Woolhiser [3], i.e., Eqs. (4-51) and (4-52). A schematic representation of the model is shown in Fig. 6-1 where infiltration and percolation are considered to be one-dimensional in the vertical direction. The model was applied to a 40 ft laboratory-scale watershed, a field plot, and to several small rangeland watersheds.

Application of Model to Laboratory-Scale Watershed

A schematic of the laboratory physical model is shown in Fig. 6-2. The flume was filled with Poudre fine sand and the surface was covered with gauze to prevent splash erosion. Runoff was measured from the single uniform plane and gamma ray attenuation was employed to follow the vertical movement of soil moisture. The desorption portion of the moisture characteristic and the conductivity-tension relation for the soil were determined experimentally and placed in the form suggested by Brooks and Corey [4].

$$\theta \sim \psi^{\lambda} \tag{6-1}$$

and

$$K \sim \psi^{\eta} \tag{6-2}$$

where θ is water content, ψ is suction and K is conductivity.

The overland flow was considered to be laminar since the exponent m in Eq. (4-1) was experimentally determined to be equal to 2.

The results of their experiments with mechanically simulated rainfall are shown in Fig. 6-3 and 6-4. The mathematical model closely represented these two hydrographs generated in the laboratory watershed. The two figures represent hydrographs simulated under initially dry and wet conditions and the runs were essentially at equilibrium.

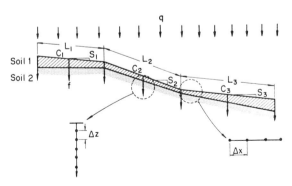

FIG. 6-1. Schematic representation of the mathematical watershed model. (After Smith and Woolhiser, *Water Resources Res.* **7**, No. 4, 899–913 (Aug. 1971), copyright by American Geophysical Union [2].)

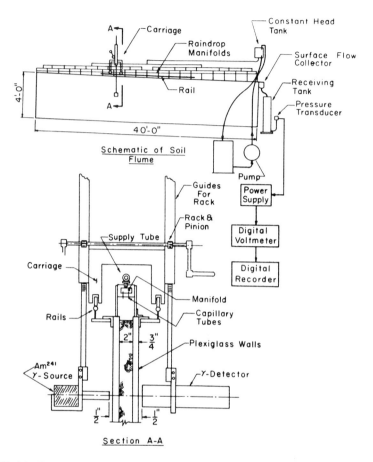

FIG. 6-2. General schematic of the laboratory soil flume and instrumentation for studying watershed response. (After Smith and Woolhiser, *Water Resources Res.* **7**, No. 4, 899–913 (Aug. 1971), copyright by American Geophysical Union [2].)

Application of Model to a Small Watershed Plot

The same model was applied to a small watershed plot at the Hastings, Nebraska watershed installation operated by the US Agricultural Research Service [2]. Much of the soil and hydraulic data were available from a detailed survey performed by Holtan *et al.* [5].

The soil type was Colby silt loam and the plot was an unfurrowed natural pasture 300 ft long and 100 ft wide in native grass vegetation. Natural rainfall was measured by a continuous recorder and the intensities

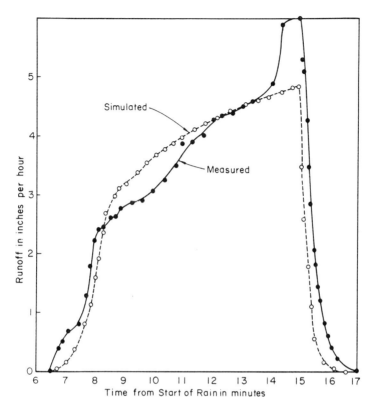

FIG. 6-3. Measured and simulated hydrographs for run 4 with moist initial conditions. (After Smith and Woolhiser, *Water Resources Res.* **7**, No. 4, 899–913 (Aug. 1971), copyright by American Geophysical Union [2].)

were determined at 1-min intervals. The soil moisture was measured bi-annually at 1-ft increments to a depth of 4 ft by a volumetric sampling.

The results of the model application to three storms are shown in Figs. 6-5 to 6-7. Each computer simulation used approximately 1 sec of the CDC 6400 for each minute of simulated storm and approximately 70,000 core storage.

In this example, Smith and Woolhiser [2] had to adjust the model parameters to arrive at the best fit in each of the three storms. This was necessary because the soil properties had to be estimated, whereas in the laboratory watershed, all properties were measured directly. Hence, as mentioned in Chapter 1, there would be a need at this point for an objective best fit criteria. However, the model did provide a theoretical framework

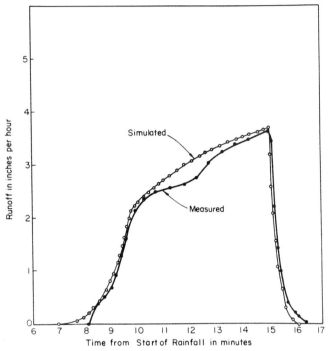

FIG. 6-4. Measured and simulated hydrographs for run 3 with dry initial conditions. (After Smith and Woolhiser, *Water Resources Res.* **7,** No. 4, 899–913 (Aug. 1971), copyright by American Geophysical Union [2].)

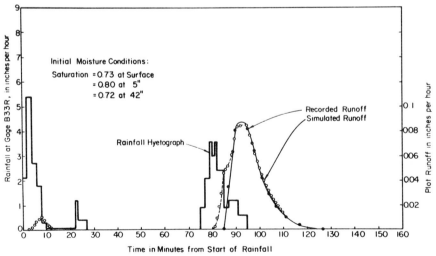

FIG. 6-5. Rainfall pattern, measured and simulated hydrographs for June 17, 1944, storm on pasture plot 56-H at Hastings, Nebraska. (After Smith and Woolhiser [2].)

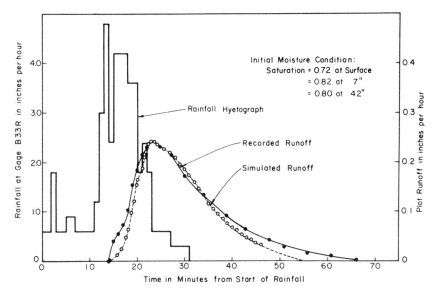

FIG. 6-6. Rainfall pattern, measured and simulated hydrographs for June 29, 1944, storm on pasture plot 56-H at Hastings, Nebraska. (After Smith and Woolhiser [2].)

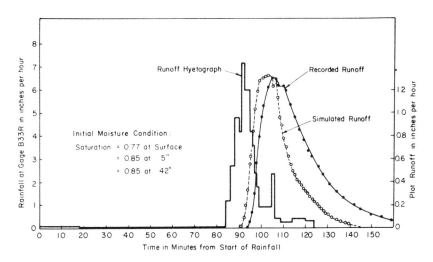

FIG. 6-7. Rainfall pattern, measured and simulated hydrographs for June 5, 1945, storm on pasture plot 56-H at Hastings, Nebraska. (After Smith and Woolhiser [2].)

for a good description of the watershed response in which system parameters with physical significance may be obtained by comparison with experimental data.

Application of Model to Rangeland Watersheds

Woolhiser *et al.* [6] applied the kinematic model to several small rangeland watersheds at Cottonwood, South Dakota. Geometry of the watersheds was approximated by a series of planes and channels as shown in Fig. 6-8. These watersheds were each about 2 acres in size, and storms chosen were long and intense enough to assume a constant infiltration rate during the last hour. The infiltration rates were estimated assuming that surface storage at the beginning of the intense portion of the storm was the same as the storage remaining at the same flowrate during the recession. The infiltration during this interval was assumed to be equal to the difference between rainfall and storm runoff. These infiltration rates averaged out to be about 0.10 in./hr.

Recognizing that much of the system parameters could not be measured or were unavailable, the need for an objective fitting function was clearly indicated. A search technique was developed to optimize the model parameters by minimizing the objective function

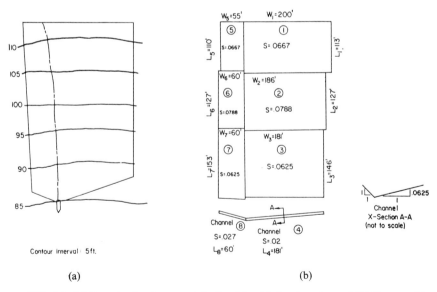

(a)   (b)

FIG. 6-8. (a) Cascade of *n* planes contributing lateral inflow to a channel. (b) Topographic map and cascade representation of watershed H-3, Cottonwood, South Dakota. (scale, 1:2850.) (After Woolhiser *et al.* [6].)

$$F = \sum_{i=1}^{n} (Q_{obs} - Q_{model})^2 \tag{6-3}$$

The model was applied to both lightly and heavily grazed watersheds.

The watersheds varied from approximately 400 to 600 ft in length, but even with these lengths, they concluded that turbulent flow did not occur on the lower planes of the cascade. As in the laboratory and field plot examples, the general laminar relation used by Morgali [7] was incorporated in the model.

$$C = \sqrt{\frac{8gN_R}{K}} \tag{6-4}$$

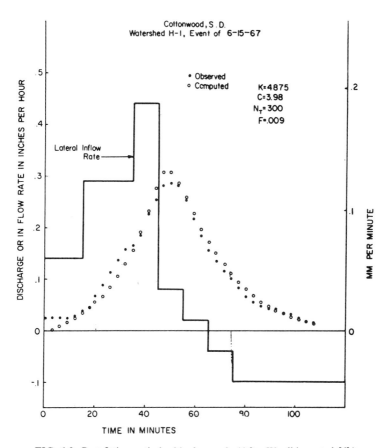

FIG. 6-9. Best-fitting optimized hydrograph. (After Woolhiser *et al.* [6].)

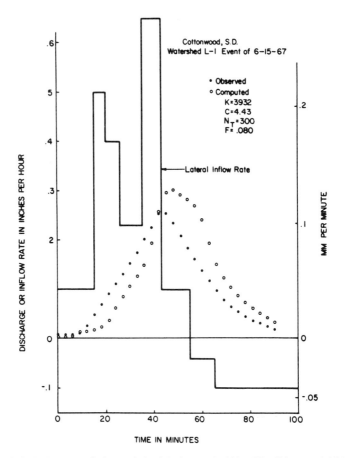

FIG. 6-10. Worst-fitting optimized hydrograph. (After Woolhiser *et al.* [6].)

where C is the Chezy resistance coefficient, N_R the Reynolds number, and K a coefficient related to geometry and vegetation. The coefficient K was optimized for storms on four of the experimental watersheds with an average error of about 5%. Examples of the best and worst fits are shown in Fig. 6-9 and 6-10, where N_T is optimized transition Reynolds number.

Woolhiser *et al.* [6] tested the predictive capability of the model by calculating hydrographs on four of the moderately grazed watersheds (not used in the analysis) for the same storm analyzed on watersheds L-1 and H-1 using the mean of the optimized K-values for the lightly and heavily grazed watersheds. Examples of the best and worst fitting hydrographs are shown in Fig. 6-11 and 6-12.

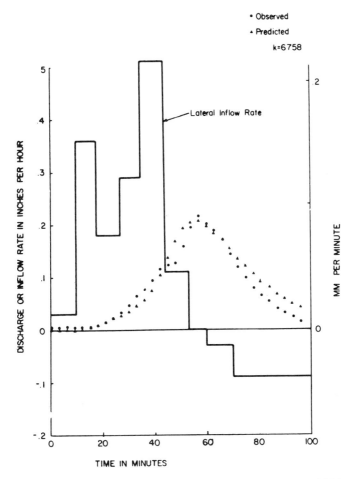

FIG. 6-11. Best-fitting predicted hydrograph. (After Woolhiser *et al.* [6].)

It is important to examine the sources of error in this experiment. The synchronization errors associated with the rainfall and runoff clocks could be as large as ±10 min, and the variability in the K-values for the heavily grazed watershed could easily be accounted for by synchronization errors [6]. This synchronization error was thought to be very large relative to the errors associated with the assumption of a constant infiltration rate and the possible errors in measuring rainfall and runoff intensity.

It was concluded that the kinematic cascade model accurately described the stormwater response under conditions of predominantly overland flow. The roughness parameter K was found to be about 7000 for short-grass

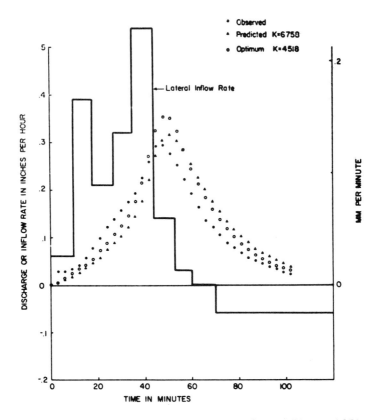

FIG. 6-12. Worst-fitting predicted hydrograph. (After Woolhiser *et al.* [6].)

prairie watersheds and the transitional Reynolds number was about 300. Also, these Reynolds numbers were consistent with those obtained by analysis of laboratory data [2].

6-3   A Model of Urban Peak Runoff Design

Overton and Tsay [8] developed a model for peak runoff design in urban areas utilizing the kinematic wave approximation to surface runoff and the SCS method for evaluating effective rainfall. The method is based upon similar concepts which underlie the rational method. For a given return period of rainfall, the time of concentration for the small watershed is set equal to the duration of the storm. By doing this, equilibrium will have been reached and the maximum peak condition would have been designed for the given return period.

For a given return period of rainfall, rainfall intensity varies inversely with the duration of the storm such that

$$i = \frac{a}{(b + D)^c} \qquad (6\text{-}5)$$

where D is rainfall duration, and a, b, and c vary with return period and geography. As an example, rainfall intensities for various return periods for Knoxville, Tennessee are shown in Fig. 6-13 [9].

From the principles developed in Chapter 5, time of concentration for a pervious catchment or watershed would be

$$t_c = \mu(i_e)^{-0.4} \qquad (6\text{-}6)$$

where μ is the lag modulus for the surface water system and i_e is the average effective rainfall intensity. If rainfall duration is set equal to time of concentration in Eq. (6-5), the result is that Eqs. (6-5) and (6-6) must be solved simultaneously. An example of this simultaneous solution is shown in Fig. 6-14. This is the solution for a hypothetical catchment in Knoxville, Tennessee with a curve number of 90 for a 10-yr return period.

Equation (6-6) can be placed in terms of the SCS curve numbers. The effective rain intensity is equal to the total volume of storm or direct runoff divided by storm duration. Using Eqs. (2-47) and (2-48),

$$i_e = \frac{60}{D} \frac{[P - (200/CN) + 2]^2}{[P + (800/CN) - 8]} \qquad (6\text{-}7)$$

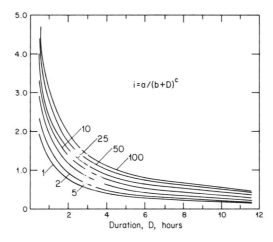

FIG. 6-13. Rainfall frequency duration relations for Knoxville, Tennessee.

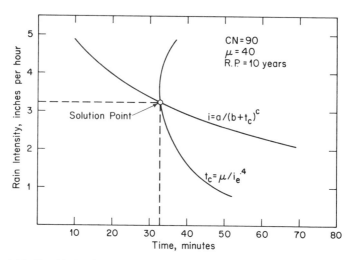

FIG. 6-14. Graphical solution for design rainfall intensity and associated time of concentration. (After Overton and Tsay [8].)

where D is in min and i_e in in./hr. Now, combining Eqs. (6-6), (6-7), and (6-5) the time of concentration equation is

$$t_c^{0.6} = \mu \left\{ \frac{60 \left[\dfrac{at_c}{60(b+t_c)^c} - \dfrac{200}{CN} + 2 \right]^2}{\left[\dfrac{at_c}{60(b+t_c)^c} + \dfrac{800}{CN} - 8 \right]} \right\}^{-0.4} \tag{6-8}$$

And the peak runoff rate in cubic feet per second is

$$Q_p = i_e \cdot DA \tag{6-9}$$

where DA is the surface drainage area in acres.

The solution procedure is to solve for time of concentration from Eq. (6-8), which is the same as the graphical solution depicted in Fig. 6-14, calculate i_e in Eq. (6-7) by setting $D = t_c$, and then calculate the peak rate in Eq. (6-9). An equivalent rational "C" can be calculated as

$$C = i_e / i_D \tag{6-10}$$

where i_D would be the associated design rainfall from Eq. (6-5). It is apparent that the rational coefficient is a function of curve number (soils, land use, antecedent moisture), and has a return period associated with it.

The solution technique for Knoxville, Tennessee was computerized by first fitting Eq. (6-5) to the rainfall-duration-frequency data shown in Fig.

TABLE 6-1. Optimized values for the parameters a, b, and c for Knoxville, Tennessee

Return period (yr)	time (min) $i = a/(b + t_c)^c$ Parameter			Coefficient of determination	Standard error (in./hr)
	a	b	c		
1	147.9	34.6	1.03	0.998	0.011
2	114.7	24.3	0.97	0.998	0.015
5	229.7	34.1	1.04	0.996	0.020
10	243.0	35.6	1.02	0.998	0.019
25	208.0	25.7	0.98	0.997	0.024
50	205.1	24.4	0.96	0.998	0.021
100	360.6	34.3	1.03	0.997	0.031

6-13 using nonlinear least squares [10]. The results of the fitting are shown in Table 6-1. The fits were all excellent and the exponent c was very nearly equal to unity for all return periods.

The Newton–Raphson method of successive approximation was used to solve Eq. (6-8). A computer program was used to generate the values of

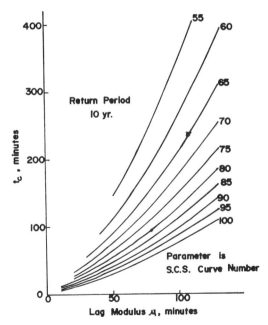

FIG. 6-15. Lag modulus versus time of concentration for catchments and watersheds in Knoxville, Tennessee. (After Overton and Tsay [8].)

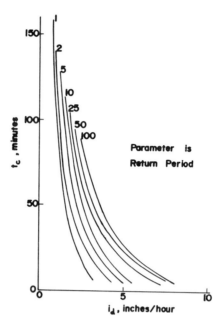

FIG. 6-16. Design rainfall intensity versus time of concentration for catchments and watersheds in Knoxville, Tennessee. (After Overton and Tsay [8].)

time of concentration t_c, design rainfall intensity i_D, and peak flow at equilibrium Q_p for return periods of 1, 2, 5, 10, 25, 50, and 100 year, with the given curve number CN, lag modulus μ, and parameters a, b, and c from Table 6-1. The results for the 10-yr return period are shown in Figs. 6-15–6-17.

Given the lag modulus, the time of concentration can be found from Fig. 6-15. The time of concentration increases with lag modulus but decreases with curve number for a given return period. The time of concentration is sensitive to low CN values. The design rainfall intensity can be found in Fig. 6-16 as a function of lag modulus and increases with CN.

In Fig. 6-17, peak flow (in./hr) can be found directly as a function of lag modulus, return period, and curve number. With these graphs the effects of land use changes on peak flow can be easily found for land use before and after development.

EXAMPLE 6.1   Assume that a pasture of 150 acres with a CN of 70 is to be developed into an airport. The 10-yr return period is chosen by the designer. Assume further that the CN after development would be 95 and lag moduli before and after development were equal to 50 min and 30 min, respectively.

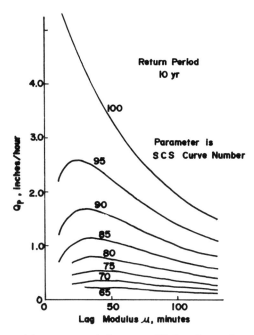

FIG. 6-17. Lag modulus versus peak flow at equilibrium for catchments and watersheds in Knoxville, Tennessee. (After Overton and Tsay [8].)

Entering Fig. 6-17 we find a peak flow of 0.37 in./hr for a CN of 70 before development and a peak flow of 2.55 in./hr for a CN of 95 after development. The increase in peak flow would be

$$(2.55-0.37) \text{ in./hr} \times 150 = 327 \text{ cfs}$$

or

$$(2.18/0.37) \times 100\% = 589\%$$

In those figures, the uppermost curve ($CN = 100$) corresponds to a 100% runoff condition. The peak discharge (in./hr) is always decreasing, while the lag modulus is increasing. However, for other curve numbers, peak flow increases initially with lag modulus until it reaches a maximum and then declines. Two reasons can be explained for this phenomenon:

(1) For very small catchments which correspond to small values of lag modulus, time of concentration is very small and corresponds to a small rainfall volume which has a large percentage lost to initial abstraction. As the size of the catchments increases, the initial abstraction accounts for a lesser percentage of total rainfall and peak runoff increases.

TABLE 6-2   Comparison of the rational and University of Tennessee peak runoff methods

Rational method	University of Tennessee method (kinematic wave using SCS curve number)
(1) It treats the surface runoff process as a linear system.	(1) It treats the surface runoff process as a nonlinear system.
(2) It is not based on applied physics.	(2) It is based on applied physics.
(3) The runoff coefficient is not considered to be a function of soils.	(3) The runoff coefficient is logically a function of soils and land use.
(4) It has no mathematical derivation and also has no experimental verification.	(4) It has mathematical derivation and the method has been verified considerably.
(5) Considerable judgment is involved in the estimation of "C" and "t_c."	(5) Little judgment is involved in estimating parameters.
(6) t_c is not considered to be a function of land use and rain rate.	(6) t_c is a function of land use and rain rate.

(2)   Since the marginal direct runoff decreases with lag modulus, although the marginal time of concentration increases, the marginal peak flow always decreases from positive to negative with lag modulus. When the marginal peak flow is equal to zero, the peak flow (in./hr) reaches a maximum.

The comparison of the Rational and the University of Tennessee [8] methods is shown in Table 6-2. By using the University of Tennessee [8] method, the runoff coefficient is logically a function of soils and land use and time of concentration is a function of land use and rainfall intensity. The rational method treats the surface runoff process as a linear system; however, the process is widely known to be very nonlinear. It must be kept in mind that the University of Tennessee method should be used only in small watersheds or catchments because it is subject to the assumptions that rainfall intensity is uniform and steady.

6-4   Use of Stormwater Detention Basins for Peak Runoff Reduction

As shown in Example 6.1, the effects of urbanization or development of pasture land into imperviousness has the effect of drastically increasing peak runoff rates. This is a matter of general knowledge, but only recently have efforts been made in urban areas to institute management methods to offset the effects of development on peak flows. Stormwater detention basins have been used recently to manage stormwater. A stormwater detention basin could amount to little more than the water backup behind a highway or road culvert.

In designing stormwater detention basins, it is necessary to route the storm hydrograph through the basin in order to size the culvert beneath the roadway. An example will be shown here of the evaluation of a road culvert which was sized to temporarily back up water behind a roadway into a small detention basin at the outlet of an 85 acre residential watershed in Knoxville, Tennessee. A schematic of the watershed is shown in Fig. 6-18. The watershed could be closely approximated as V-shaped and the Overton–Brakensiek model for the rising and falling hydrographs was assumed to be applicable. The weighted average physical and hydraulic characteristics for the basin were calculated. The following example is set out to illustrate the procedure for evaluating the effects of development on stormwater runoff and to demonstrate an easy and effective mechanism for offsetting those effects [11].

EXAMPLE 6.2   Prior to 1950, the subdivision in Fig. 6-18 was 50% pasture, 45% woods, and 5% dirt roads. The hydrologic soil type is "C" and the land use was in poor hydrologic condition. In 1960 the 85 acres had been completely developed into residential zone R-2 (three houses per acre), and presently about 40% of the land is impervious (i.e., roof tops, asphalt roads, and concrete driveways).

Presently a 36-in. concrete culvert exists under the roadway shown which backs up stormwater into the cross-hatched area and forms a detention basin. There is an additional 3.15 ft of headwater above the 3 ft culvert before the flow will overtop the roadway and is associated with a maximum detention storage of 2.17 acre-ft.

(a) Evaluate the increase in peak flow due to the development of the 85 acres associated with a rainfall event with a 100-yr return period; and

(b) evaluate the effectiveness of the existing stormwater detention basin.

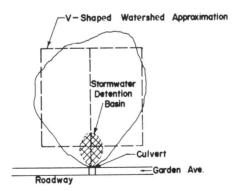

FIG. 6-18. An 85-acre residential watershed in the Harrill Heights Subdivision, Knoxville, Tennessee.

(a)   The weighted average catchment parameters were found to be

Before development: $n = 0.10$, $S_0 = 0.100$, $CN = 65$
After development: $n = 0.047$, $S_0 = 0.089$, $CN = 84$

The parameters for the V-shaped watershed approximations are

$$L_0 = 660 \text{ ft}, \qquad L_c = 2800 \text{ ft}, \qquad w = 10 \text{ ft}$$

The lag moduli before and after were

$$\mu_B = 22.8 \text{ min.}; \qquad \mu_A = 15.0 \text{ min.}$$

From the design manual of Overton and Tsay [8], the design peak flows before and after are

$$Q_p(B) = 42 \text{ cfs}, \qquad Q_p(A) = 138 \text{ cfs}$$

or an increase of 229%.
     The equivalent runoff coefficient for these two events are

$$C(B) = 0.50/4.9 = 0.102, \qquad C(A) = 1.63/7.1 = 0.229$$

(b)   In order to route the design storm through the detention basin, a rising and falling hydrograph must be hooked to the peak flow. A distinct shortcoming of conceptual peak runoff methods is that they have no hydrograph associated with them.
     The rising and falling hydrographs for the design storm were calculated using Eqs. (4-46) and (4-40), respectively.
     The next step is to develop the detention storage-discharge relation or the reservoir rating curve, as it is sometimes referred to. The detention storage-elevation curve for the detention basin was determined from a

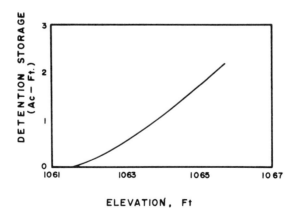

FIG. 6-19. Detention storage-elevation relation for 85 acre subdivision of Example 6.2.

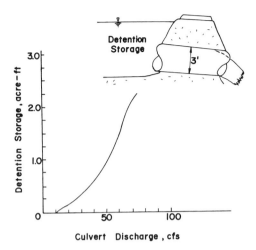

FIG. 6-20. Rating curve for stormwater detention basin in Example 6.2.

field survey and is shown in Fig. 6-19. Also shown is the culvert discharge–elevation relation for the basin. This relation was developed with the help of a hydraulic design handbook [12]. Finally, elevation is eliminated from Fig. 6-19 and the rating for the detention basin is shown in Fig. 6-20.

The storm was routed through the detention basin using the hydrologic conservation of mass equation [13].

$$I - O = \frac{dS}{dt} \tag{6-11}$$

where I is stormwater inflow to the detention basin in cubic feet per second, O the outflow from the culvert in cubic feet per second, S the detention basin storage in cubic feet per second-hours and t the time in hours. Equation (6-11) can be solved by a central difference numerical scheme as [13]

$$(I_1 + I_2) + [(2S_1/\Delta t) - O_1] = [(2S_2/\Delta t) + O_2] \tag{6-12}$$

where the subscripts 1 and 2 denote the beginning and the end of the finite time interval Δt.

The curves for $(2S/\Delta t) - O$ and for $(2S/\Delta t) + O$ as a function of outflow are shown in Fig. 6-21. The routing proceeds as shown in Table 6-3 and the results are plotted in Fig. 6-22.

The result is that the detention basin with a 36-in. culvert reduces the design peak flow from 138 cfs down to 62 using less than 2 acre-ft of storage. The estimated peak flow associated with the land use before development

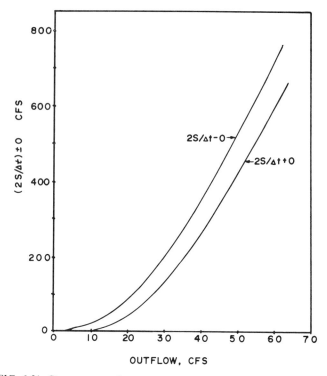

FIG. 6-21. Storage curves for stormwater detention basin in Example 6.2.

TABLE 6-3   Detention basin routing computations for Example 6.2

(1) Time (min)	(2) n	(3) I_n (cfs)	(4) $I_1 + I_2$	(5) $(2S_n/\Delta t) - O_n$ (cfs)	(6) $2S_{n+1}/\Delta t + O_{n+1}$	(7) O_{n+1} (cfs)
0	1	0	0	0		0
3	2	8	8	0	8	5.5
6	3	24	32	8	32	12
9	4	48	72	41	80	19
12	5	76	124	108	165	28
15	6	111	187	219	295	38
18	7	138	249	539	654	57
21	8	88	226	642	765	61
24	9	59	147	665	789	62
27	10	41	100	642	765	61
30	11	31	72	595	714	59
33	12	24.5	55.5	535	650	57
36	13	19.5	44	470	579	54
39	14	16	35.5	403	506	51
42	15	14	30		433	47

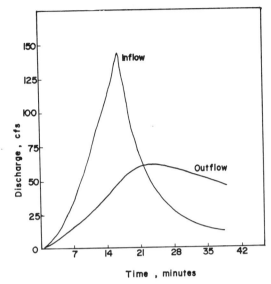

FIG. 6-22. Routing of 100-year design hydrograph through stormwater detention basin in Example 6.2. (After Environ. Planning Engr. [11].)

for the 100-yr return period was 42 cfs. Hence, the detention basin has reduced the increase in peak flow from 229% down to about 50%.

A smaller culvert size might reduce the peak flow further, but the risk of overtopping the road would be increased. However, this exercise could be repeated for a 30-in. culvert.

6.5 Hydrologic Impact of Storm Sewers

In the development of shopping centers, it has been traditional to design and construct sewer systems which would allow for a minimum of surface storage on a parking lot by quickly collecting and discharging the stormwater away from the property. Such action has generated numerous lawsuits nationwide by downstream owners who have protested the collection and concentration of upstream water and "dumping" it on the downstream owners. Basically, this is in violation of most state drainage laws. An alternative course of action would be not to construct sewers at shopping centers. Then the hydrologic impact to downstream owners could be minimized by allowing the stormwater to flow off the entire surface area, be collected in an open channel, and then utilize a small detention basin as in Example 6.2, to reduce the peak flow to what it would have been before development.

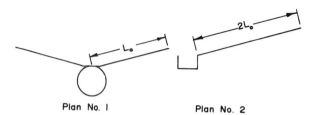

Plan No. I   Plan No. 2   FIG. 6-23.

EXAMPLE 6.3   Consider a 30-acre pasture in the headwaters of an un-developed creek near Knoxville, Tennessee. The area is to be made entirely impervious by construction of a shopping center. Evaluate the effects on the stormwater peak flow associated with a 10-yr return period for two drainage plans shown in Fig. 6-23. The first plan is to construct a 72-in. storm sewer beneath the shopping center in the location of the existing stream. The second plan is to have no sewers, but to slope the parking lot such that all water collects in an open ditch. Neglect t_c for open channel flow. The following data are available:

Before development: $CN = 65$, $n_0 = 0.10$, $L_0 = 1000$ ft, $S_0 = 0.01$
  After development: $CN = 100$, $n_0 = 0.25$,   and $S_0 = 0.01$

Before development, the lag modulus is 58.5, and from Fig. 6-17 the peak flow is 0.20 in./hr or 6 cfs.

   For Plan (a), the lag modulus is 16.8 and from Fig. 6-17, the peak flow is 5.1 in./hr or 153 cfs. This is an increase of 2550%! For Plan (b), the lag modulus is 25.5 and from Fig. 6-17, the peak flow is 4.5 in./hr or 135 cfs. This is an increase of 2250%! Plan (b) allows for less increase in peak flow but it also results in cost savings in storm sewers plus a detention basin can be designed for construction on the property to allow for offsetting the peak flow associated with development. If a detention basin were con-structed with Plan (a), backup in the piping system could occur.

6-6   Stormwater Management Models

Combined Sewer Overflows

   In 1971, a model capable of simulating urban stormwater runoff and combined sewage overflows was developed by Water Resources Engineers, Inc., Metcalf and Eddy, Inc., and the University of Florida [14] for the Environmental Protection Agency. This EPA model takes rainfall and basin characteristics as input and calculates stormwater quantity and quality.
   As illustrated in Fig. 6-24, the model consists of four components;

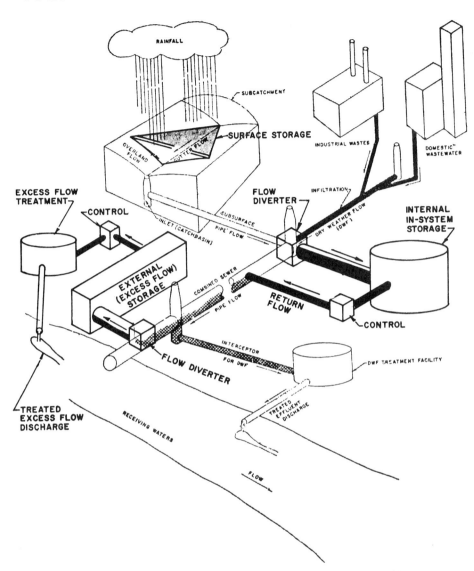

FIG. 6-24. EPA stormwater management model. (After Lager *et al.* [14].)

runoff, transport, storage, and receiving water. The runoff component computes direct runoff for a given storm for each subcatchment using an empirical regression equation developed at Johns Hopkins University [15]. This becomes the input at the inlets to the main sewer system. The transport component routes the flows through the sewer systems using a form of the kinematic flow approximation, i.e., Manning's equation and conservation of mass. The storage component modifies the sewer flow in the system at points where facilities are provided. At this point, costs associated with required storage for various return periods are calculated. Finally, the receiving component computes the effects of the sewer discharge in the river, lake, or bay.

The EPA model has been widely used in the United States in consulting ventures by Water Resources Engineers, Inc. Examples of application of the model to water quality management problems will be discussed in Chapter 16.

Effects of Urbanization

Bras and Perkins [16] utilized the MIT model [17] to quantify the likely effects of typical urban developments on the hydrologic response of hypothetical catchments and to relate these effects to the separate physical

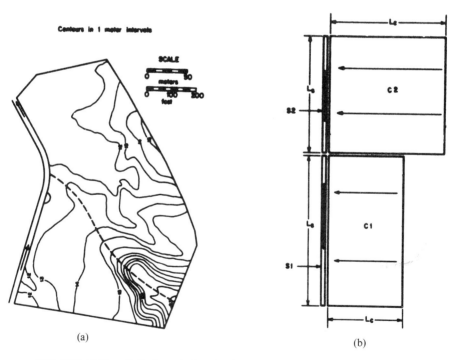

(a)   (b)

FIG. 6-25 (a) Hypothetical area before development. (elevation contours shown). (b) Four-element schematization of hypothetical area before development. (After Bras and Perkins [16].)

TABLE 6-4 Data for computer simulation of simple representation of hypothetical catchment after development[a]

Element		S1	S2	C1	C2
Flow length L (ft)	(1)	515	312	625	834
Effective flow length L^* (ft)[b]	(2)	—	—	343.5	1050
Width (ft)	(3)	—	—	515	312
Slope	(4)	0.007	0.007	0.035	0.022
Manning's n	(5)	0.10	0.10	0.087	0.087
α	(6)	0.526	0.526	3.20	2.54
α^*	(7)	—	—	1.76	3.20
m	(8)[c]	1.33	1.33	1.67	1.67

[a] After Bras and Perkins [16].   Note: 1 ft = 0.305 m.
[b] See ref. 2 in Bras and Perkins [16].
[c] α^* is the adjusted value $= \alpha(L^*/L)$.

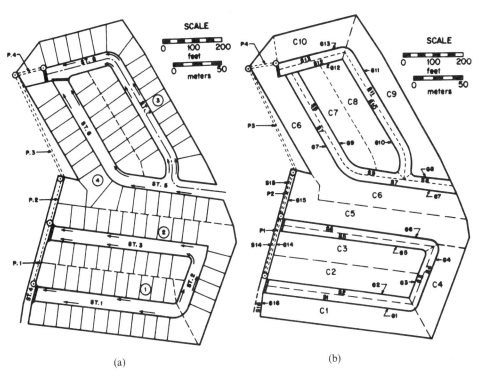

(a)   (b)

FIG. 6-26. (a) Hypothetical area after development showing lot, street, inlet, and sewer arrangement. (b) Detailed segmentation of hypothetical area. (After Bras and Perkins [16].)

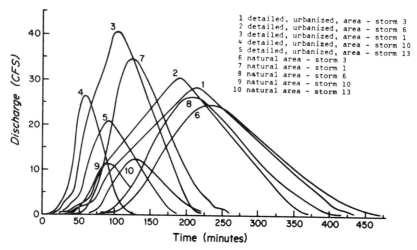

FIG. 6-27. Hydrographs from detailed and natural representation of hypothetical area with Horton's type infiltration and roofs discharging into surrounding pervious areas. (After Bras and Perkins [16].)

changes introduced by urbanization. The MIT model uses Horton's model to compute infiltration and the kinematic wave approximation for computing overland and open channel flow.

A catchment area in Puerto Rico was hypothesized as shown in Fig. 6-25. The natural catchment was subdivided into two overland segments and two river segments. The catchment was 11.5 acres and the natural characteristics are shown in Table 6-4.

The hypothetical catchment was assumed to be developed into a residential area with lots and streets as shown in Fig. 6-26. Nine hypothetical storm hyetographs were used, eight were symmetrical triangular storms and the ninth storm hyetograph was a square. The results are shown in Fig. 6-27.

The principal results of the MIT study were:

(1)   Urbanization increased peak discharge from 7 to 200% and time to peaks were reduced from 8 to 40%.

(2)   The magnitude of the urbanization effect depended upon catchment as well as rainfall characteristics. This finding is consistent with the kinematic wave theory detailed in Chapter 4.

(3)   Infiltration was found to be important but changes in catchment response were observed even when no infiltration was permitted.

(4)   Effects of rooftop storage were very small but if properly designed they could be used to counteract urbanization very effectively, and

(5)   The spatial distribution of developed areas in the basin has an important and sensitive effect on hydrologic response.

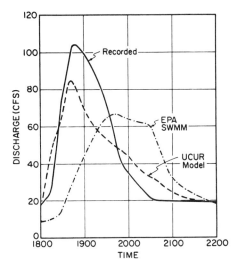

FIG. 6-28. Calculated and recorded runoff hydrographs storm of Nov. 9, 1970—outlet of Bloody Run Sewer Watershed, Cincinnati, Ohio. (After Papadakis and Preul [18].)

6-7 A Comparison of Urban Stormwater Models

Papadakis and Preul [18] performed a comparison of urban runoff models using design and observed rainstorms for several catchments and

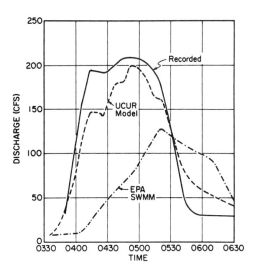

FIG. 6-29. Calculated and recorded runoff hydrographs storm of May 13, 1971—outlet of Bloody Run Sewer Watershed, Cincinnati, Ohio. (After Papadakis and Preul [18].)

watersheds in Chicago, Illinois, and Cincinnati, Ohio. All the models tested are fundamentally based upon the kinematic approximation for routing flows overland and in open channels.

All the methods tested successfully reproduced stormwater runoff hydrographs for the small drainage areas. Comparisons were made between the recorded storm hydrographs and the EPA and University of Cincinnati model [19] (UCUR) for the Bloody Run Sewer Watershed in Cincinnati. It has a surface drainage area of 2380 acres drained by a combined sewer system designed for a 10-yr return period. The area has a population density of 12 persons/acre and 45% of the area is impervious. The results of the storm simulations are shown in Figs. 6-28–6-30.

The differences in these comparisons indicated that the estimation of the infiltration parameters has a very sensitive effect upon simulated stormwater hydrographs for large urban watersheds. These results, coupled with the results of the MIT model applied to the Puerto Rico catchment indicate that small urban watersheds are very flashy, and infiltration estimates therefore do not have a very sensitive effect upon stormwater simulations relative to larger urban areas. As drainage area increases, the watershed performs more of a dampening effect upon stormwater by way of the open channel flow routings. Since hydrologic response is more sluggish, infiltration estimates play an increasingly important role in stormwater simulations.

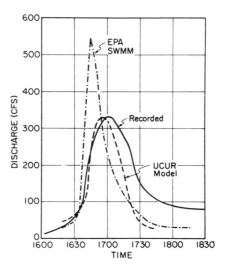

FIG. 6-30. Calculated and recorded runoff hydrographs storm of Aug. 25, 1971—outlet of Bloody Run Sewer Watershed, Cincinnati, Ohio. (After Papadakis and Preul [18].)

Problems

6-1. What is the mathematical structure of the Smith–Woolhiser model? How much confidence do you place on it relative to the EPA or MIT models?

6-2. Contrast the basic input data requirements of each of the models presented in this chapter.

6-3. Estimate the peak flow (in cfs) with a 10-yr return period for the watershed in Example 5.2 if the land use is pasture in poor hydrologic condition and has a type C soil. The watershed is near Knoxville, Tennessee.

6-4. Estimate the rising and falling limbs of the hydrograph associated with the design peak flow in Problem 3.

6-5. If the watershed in Problem 3 were completely paved with asphalt without modifying the overland slope or the channel conditions, estimate the percent increase in the 10-yr peak flow and time of concentration.

6-6. Calculate the "rational coefficients" associated with Problem 5. Contrast the results with a table of "rational coefficients" you may have handy.

6-7. Repeat Example 6.2 if the weighted average roughness of the catchment was $n = 0.035$ and the road culvert was 4 ft in diameter.

6-8. What were the main conclusions which Bras and Perkins arrived at concerning the effects of rooftop storage and land use distributions on stormwater in their study catchment?

6-9. On the basis of the results reported by Papadakis and Preul, which of the urban stormwater models tested do you believe is the most reliable? Explain.

References

1. American Society of Civil Engineers, Urban water resources research, A Report to the Office of Water Resources Research, US Dept. of Interior, New York, Sept. 1968.
2. Smith, R. E., and Woolhiser, D. A., Overland Flow on an Infiltrating Surface, *Water Resources Res.* **7**, No. 4, 899–913 (Aug. 1971).
3. Kibler, D. F., and Woolhiser, D. A., The kinematic cascade as a hydrologic model, *Colorado State Univ. Hydrology Paper*, No. 9, (1970).
4. Brooks, R. H., and Corey, A. T., Hydraulic properties of porous media, *Colorado State Univ. Hydrology Paper*, No. 3 (March 1964).
5. Holtan, H. N. *et al.*, Moisture-tension data for selected soils on experimental watersheds, ARS 41-144, US Dept. Agr., Agr. Res. Service, Beltsville, Maryland, Oct. 1968.
6. Woolhiser, D. A., Hanson, C. L., and Kuhlman, A. R., Overland flow on rangeland watersheds, *J. Hydrol.* (*New Zealand*) **9**, No. 2, 336–356 (1970).
7. Morgali, J. R., Laminar and turbulent overland flow, *J. Hydr. Div., Amer. Soc. Civil Engr.* **96**(HY2), 441–460 (1970).

8. Overton, D. E., and Tsay, C. G., "A kinematic method of urban peak runoff design using SCS curve numbers for Knoxville, Tennessee," Dept. of Civil Eng., Res. Series No. 25, Knoxville, Tennessee, Aug. 1974.

9. Hershfield, D. M., Rainfall frequency atlas of the United States, US Weather Bureau Tech. Paper No. 40, 1961.

10. DeCoursey, D. G., and Snyder, W. M., A computer-oriented method of optimizing hydrologic model parameters, J. Hydrol. 9, 34–56 (1969).

11. Environmental Planning Engineers, Inc., Least cost design of drainage facilities in Harrill Heights Subdivision, A Design Report Submitted to the City of Knoxville, Tennessee, Dec. 1974.

12. Davis, C. V., "Handbook of Applied Hydraulics," 2nd ed. McGraw–Hill, New York, 1952.

13. Linsley, R. K., Kohler, M. A., and Paulhus, J. L. H., "Applied Hydrology." McGraw–Hill, New York, 1949.

14. Lager, J. A., Shubinski, R. P., and Russell, L. W., Development of a simulation model for stormwater management, J. Water Pollution Control Fed. 43, 2424–2435 (Dec. 1971).

15. Geyer, J. C., and Lentz, J. A., An evaluation of the problems of sanitary sewer design, Final Report of the Residential Sewerage Research Project, Dept. of Environ. Eng. Science, Johns Hopkins Univ., 1963.

16. Bras, R. L., and Perkins, F. E., Effects of urbanization on catchment response, J. Hydr. Div., Amer. Soc. Civil Engr. 101(HY3), 451–466 (March 1975).

17. Bras, R. L., Simulation of the effects of urbanization on catchment response, thesis presented to M.I.T., Cambridge, Massachusetts, 1973.

18. Papadakis, C. N., and Preul, H. C., Testing of methods for determination of urban runoff, J. Hydr. Div., Amer. Soc. Civil Engr. 99(HY9), 1319–1335 (Sept. 1973).

19. Papadakis, C. N., and Preul, H. C., University of Cincinnati urban runoff model, J. Hydr. Div., Amer. Soc. Civil Engr. 98(HY10), 1789–1804 (Oct. 1972).

Chapter 7 | Systems Approach to Deterministic Stormwater Modeling

7-1 Introduction

A model is presented in this chapter which permits the direct calculation of hydrographs for catchments and watersheds from rainfall excess. It is an input–output approach whereby a variable response function, derived from the data of the Corps of Engineers [1] is convoluted with each period of constant rainfall excess and then summed to form the total stormwater hydrograph. The main advantages of this variable response model (VRM) are that it has a considerably lighter computational load than the load associated with the solution of the characteristic equations shown in the previous chapters, and the general complexity of the use is relatively simple. The comparisons of the VRM with the solutions of the kinematic wave characteristic equations are shown. The two models generate approximately the same output.

Prior to development and application of the VRM, it will be necessary to introduce some fundamentals of linear and nonlinear hydrologic systems.

7-2 Linear and Nonlinear Hydrologic Systems

As pointed out in Chapter 1, linear and nonlinear systems are defined by linear and nonlinear differential equations, respectively. In hydrologic terms this means that the response from a catchment watershed will remain constant for a prescribed set of boundary conditions. The response of a watershed is represented as a unit hydrograph [2]; hence, for a specified duration of rainfall D the unit hydrograph $U(D, t)$ is found by dividing the storm hydrograph ordinates by the associated volume of rainfall excess P_e. This forms a new hydrograph, i.e., the unit hydrograph, with 1 in. of runoff volume beneath it.

The volume under the storm hydrograph (see Fig. 7-1) is

$$P_e = \int_0^\infty Q(t)\, dt = i_e \cdot D \qquad (7\text{-}1)$$

where Q is stormwater discharge. The unit hydrograph then is

$$U(D, t) = Q(t)/P_e \qquad (7\text{-}2)$$

since the volume beneath it is seen to be 1 in., from Eq. (7-1).

$$1 \text{ in.} = \int_0^\infty \frac{Q(t)}{P_e}\, dt \qquad (7\text{-}3)$$

The unit hydrograph concept says that for a given land use, initial moisture content, and rainfall excess duration, the unit hydrograph will be the same for each storm. Much evidence has been reported which has shown that the unit hydrograph is also a function of rainfall excess intensity. This simply means that the system is nonlinear. The example of Minshall [3], shown in Fig. 7-2, illustrates this variation on a small agricultural watershed. These five storms have nearly the same duration of rainfall excess but have widely varying rainfall excess intensities and this resulted in the wide variation in unit hydrographs as shown in Fig. 7-2. The variation of lagtime with rain excess intensity was shown by Overton [4] and is repeated in Fig. 7-3. This result is analogous to the deterministic analysis of stormwater presented in the previous chapters. A graphical representa-

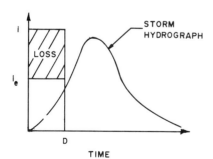

FIG. 7-1. The unit hydrograph concept.

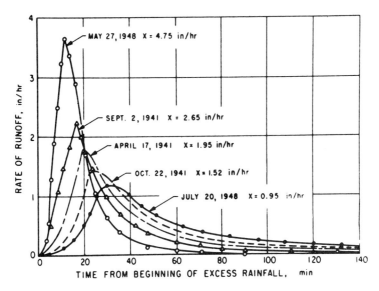

FIG. 7-2. The example of Minshall. (After Minshall [3].)

tion of the variation of lagtime with rain rate on a concrete surface of the Corps of Engineers [1] is shown in Fig. 7-4 [5]. Hence, experimental evidence of the nonlinearity of watershed response is closely correlated with the deterministic analysis and results presented in the previous chapters.

7-3 Derivation of Response Function

If 1 in. of rain excess were to be instantaneously dropped on a watershed, the stormwater hydrograph would be the unit hydrograph for a storm duration approaching zero. This is referred to as the "instantaneous unit

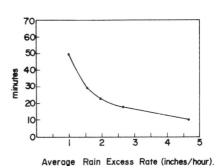

FIG. 7-3. Variation of lagtime of unit hydrographs of Minshall example. Lagtime versus supply rate. (After Overton [4].)

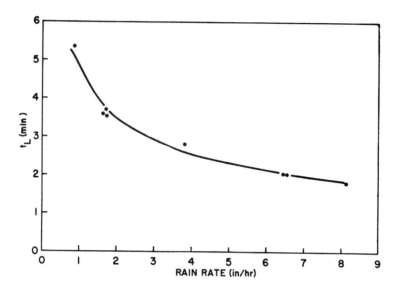

FIG. 7-4. Lagtime versus rain rate. Concrete, $\frac{1}{2}\%$ slope, 84 ft.

hydrograph" or the "instantaneous response function" (IRF) of the system. The response function can be obtained by convoluting the unit hydrograph for a finite duration D, and then taking the first derivative with respect to time. The convolution of the unit hydrograph is termed the S-curve and is equal to

$$S(t) = \int_0^t U(D, \tau) \, d\tau \tag{7-4}$$

EXAMPLE 7.1   Convolute a steady unit rainfall excess rate of 1 hr duration with the instantaneous unit hydrograph u, $u = e^{-t}$.

The solution for $0 \le t \le 1$ *hr is*

$$Q = \int_0^t i(\tau) \, e^{-t+\tau} \, d\tau \quad \text{or} \quad Q = 1 \, (\text{in./hr})(1 - e^{-t})$$

For times greater than 1 hr, the solution is

$$Q = Q(1 \text{ hr})(1 - e^{-t}) - \int_1^t e^{-t+\tau} \, d\tau \quad \text{or} \quad Q = Q(1 \text{ hr}) \, e^{-(t-1)}$$

The results are shown in Fig. 7-5.

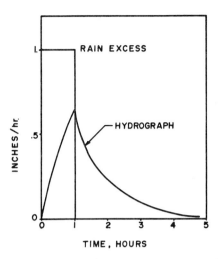

FIG. 7-5. Stormwater hydrograph derived in Example 7.1.

The derivative of the S-curve is the

$$\text{IRF}(t) = \frac{dS(t)}{dt} = \frac{d}{dt} \int_0^t U(D, \tau) \, d\tau \tag{7-5}$$

From this discussion, it should be apparent, that the observed dimensionless rising hydrograph shown in Fig. 4-3 is nothing more than an S-curve, hence an IRF for overland flow can be obtained by taking the first derivative of the function. The IRF can then be convoluted for increments of rainfall excess, lagged and summed, and thus provide a computationally efficient alternative to the solution of the kinematic wave characteristic equations.

This operation would be best achieved by first finding a mathematical expression which accurately fits the observed S-curve. None of the kinematic models shown in Fig. 4-4 would be appropriate because their first derivatives do not look like instantaneous unit hydrographs. The IRF of the Manning kinematic model would be

$$\text{IRF}_M = \begin{cases} \frac{5}{3} t_*^{2/3}, & 0 \le t_* \le 1 \\ 0, & t_* > 1 \end{cases} \tag{7-6}$$

and Eq. (7-6) is plotted in Fig. 7-6. In fact, there is a discontinuity at $t_* = 1$. To offset these problems an empirical expression of the observed S-curve was selected.

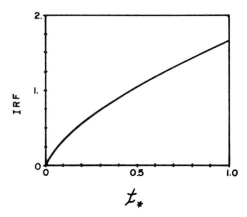

FIG. 7-6. IRF of Manning kinematic model.

The empirical expression selected was

$$Q_* = (1 - e^{-\omega T})/(1 + Ae^{-\omega T}) \qquad (7\text{-}7)$$

where

$$A = e^{1/B} - 1$$

$$\omega = (A + 1)/AB \qquad (7\text{-}8)$$

$$T = t/t_L \qquad (7\text{-}9)$$

The optimum value of B was found to be 0.20 which produced a standard error of 10% on the entire hydrograph and 15% between $0 \le T \le 1$. Values of B are plotted with the standard errors in fitting the observed S-curve in Fig. 7-7.

The response function of Eq. (7-7) for $B = 0.20$ is

$$t_L U = t_L \frac{dQ_*}{dt} = \frac{374\,Q_*^{\,2}}{\cosh(5.04T) - 1} \qquad (7\text{-}10)$$

The fitted S-curve (called log distribution) is plotted with the observed S-curve in Fig. 7-8; its associated VRF is plotted with the VRF derived from the observed S-curve in Fig. 7-9. The interpolative procedure reported by Snyder [6] was used to derive the IRF from the S-curve in Fig. 7-9.

7-4   Development of the Variable Response Model (VRM)

The VRM model computes overland flow hydrographs generated by unsteady nonuniform rainfall on hillslopes formed as a cascade of planes

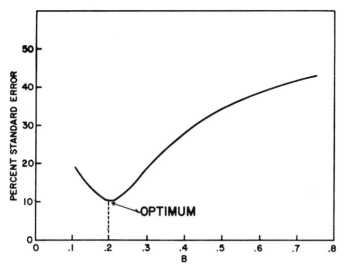

FIG. 7-7. B-value in Eq. (7-7) versus the standard error in fitting the observed S-curve in Fig. 4-3.

and for open channels with overland flow as input. This is accomplished by first transforming a cascade of planes into an equivalent uniform plane. The equivalent plane is then partitioned into sections of uniform rain intensity; hydrographs are computed for each section of the plane by convoluting the time series of rainfall with the VRF previously derived from the observed S-curve; finally these hydrographs are transmitted to the end of the plane by a storage routing function. Channel flow is computed in a

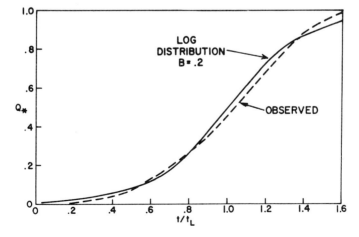

FIG. 7-8. Log distribution for optimum B-value and the observed S-curve.

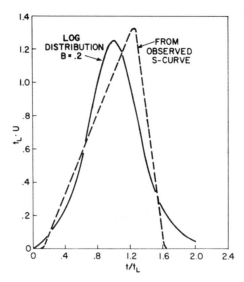

FIG. 7-9. VRF associated with log distribution and VRF derived from observed S-curve.

similar manner. The components of the model are a convolution operation, a routing function, and a technique for transforming a hillslope into an equivalent uniform plane.

Convolution

Hydrographs generated by a uniform unsteady rainfall are computed by a convolution operation. Because of the variability of lagtime with rain intensity exemplified by Fig. 7-5, a different response function was associated with the jth period of steady rain intensity of the hyetograph, $U[t, t_L(j)]$. A hydrograph for each rain period was synthesized by first computing lagtime from Eq. (4-33) and then convoluting. Hydrographs for all rain periods are then summed to form the total hydrograph.

If the duration of the first rain was applied between $t = 0$ and $t = t_1$, the second between t_1 and t_2, and the third between t_2 and t_3, the overland flow hydrograph can be represented by the convolution integral.

For $0 \leq t \leq t_1$

$$Q(t) = i(1) \int_0^t U[t - \tau, t_L(1)] \, d\tau \qquad (7\text{-}11)$$

For $t_1 \leq t \leq t_2$

$$Q(t) = i(1) \int_0^t U[t - \tau, t_L(1)] \, d\tau + i(2) \int_{t_1}^t U[t - t_1 - \tau, t_L(2)] \, d\tau \quad (7\text{-}12)$$

For $t_2 \leq t \leq t_3$

$$Q(t) = i(1) \int_0^t U[t - \tau, t_L(1)] \, d\tau + i(2) \int_{t_1}^t U[t - t_1 - \tau, t_L(2)] \, d\tau$$

$$+ \; i(3) \int_{t_2}^t U[t - t_2 - \tau, t_L(3)] \, d\tau \quad (7\text{-}13)$$

Lagtimes are computed by Eq. (4-33).

Carrying this line of thought further, a general expression for the overland flow hydrograph can be developed for any rain storm letting $i(j)$ represent the steady uniform rain intensity for the jth time interval

$$Q(t) = \sum_{j=1}^k i(j) \int_{t_{j-1}}^t U[t - t_j - \tau, t_L(j)] \, d\tau \quad (7\text{-}14)$$

The simulated overland flow hydrograph can be placed in terms of S-curves by integrating Eq. (7-14):

$$Q(t) = \sum_{j=1}^k i(j)\{Q_*[t - t_{j-1}, t_L(j - 1)] - Q_*[t - t_j, t_L(j)]\} \quad (7\text{-}15)$$

If the storm is tabulated at equal time intervals Δt, Eq. (7-15) can be discretized as

$$Q(k \, \Delta t) = \sum_{j=1}^k i(j)\{Q_*[(k + 1 - j) \, \Delta t, t_L(j - 1)]$$

$$- \; Q_*[(k - j) \, \Delta t, t_L(j)]\} \quad (7\text{-}16)$$

Routing Function

A routing procedure was developed to lag and attenuate rain that falls only on the upper section of the slope. Consider the uniform plane in Fig. 7-10. If rain is falling at a steady uniform rate of q cfs per foot length per foot width over the entire length, the total hydrograph at the end of the plane could be considered as the sum of

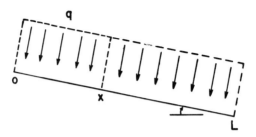

FIG. 7-10. Derivation of the routing function for a uniform plane.

1. the hydrograph produced at L due to the rainfall on the plane between x and L, $Q[L - x]$, plus

2. the hydrograph routed from x to L due to rainfall over the plane between O and x, $Q[x/L]$.

Therefore,

$$Q[L] = Q[L - x] + Q[x/L] \qquad (7\text{-}17)$$

In terms of S-curves, Eq. (7-17) can be written as

$$LqQ_*[L] = (L - x)qQ_*[L - x] + xqR[x] \qquad (7\text{-}18)$$

The routing function $R[x]$ is the only unknown in Eq. (7-18) and is

$$R[x] = (L/x)Q_*[L] - [(L - x)/x]Q_*[L - x] \qquad (7\text{-}19)$$

Using lagtimes for lengths L and $L - x$ calculated from Eq. (4-33), the routing function was calculated for a number of values of x/L. The results are plotted in Fig. 7-11. Unfortunately, severe negative values in the function resulted for ratios of x/L less than 0.8. Routed hydrographs over long distances were negative in the early rising portion of the hydrograph. These negative values are forced out by finding the least squares best fit for the parameter B in Eq. (7-7) subject to the constraint that the routing function must be positive. Mathematically, the optimum value of B was found by finding

$$\text{variance}[Q_* - \hat{Q}_*] = \text{minimum} \qquad (7\text{-}20)$$

subject to

$$L\hat{Q}_*[L] - (L - x)\hat{Q}_*[L - x] \geq 0 \qquad (7\text{-}21)$$

A new optimum B-value of 0.30 produced a standard error up to 17% in fitting the observed S-curve. The routing function of $B = 0.30$ is generalized in Fig. 7-12.

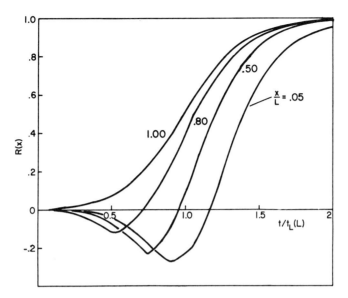

FIG. 7-11. Routing function for $B = 0.20$ for various routing lengths.

The routing function $R[x]$ is actually a dimensionless routed hydrograph over the section x to L. If it is assumed that the routing function is independent of the condition from x to L, then a hydrograph can be predicted at the end of the plot generated by rainfall on the upper section only. This will be,

$$Q[x/L] = qxR[x] \tag{7-22}$$

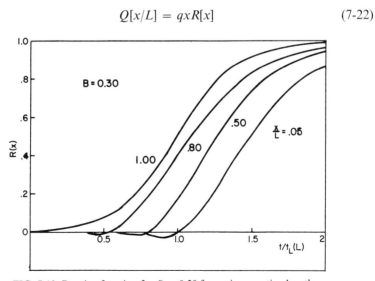

FIG. 7-12. Routing function for $B = 0.30$ for various routing lengths.

If rain rates are different on the two sections, the total hydrograph can be predicted by calculating $R[x]$ from Eq. (7-19), $Q[x/L]$ from Eq. (7-22) and adding this to the hydrograph produced at L from rain over x to L.

The assumption that the routing function is independent of the conditions between x and L may be justified using the results of Woolhiser and Liggett [7]. Since the kinematic wave accurately describes overland flow on these surfaces, the flow is approximately uniform. No significant disturbances are moving upslope; and it follows that flows under these conditions should not be significantly affected by the input over x to L. There is a great lack of information on this hydraulic process. It is known, however, that the shallow water equations are not applicable here.

Transformation of a Nonuniform Surface into an Equivalent Uniform Plane

Consider the hillslope shown in Fig. 7-13 formed as a cascade of two planes with steady uniform rainfall. Equation (4-33) relates lagtime and time of concentration to roughness, slope, and length of a uniform plane. Plane 1 can be replaced with a theoretical plane of different slope, length, and roughness so long as the new plane has the same lagtime as the actual plane. The hillslope can be replaced by a theoretically equivalent uniform plane by setting the roughness and slope of the equivalent plane 1 equal to the roughness and slope of plane 2.

$$n_1 L_1/\sqrt{S_1} = n_2 L_1'/\sqrt{S_2} \tag{7-23}$$

where L_1' is the length of the theoretically equivalent plane. The length of the equivalent plane is the only unknown in Eq. (7-23), and is found to be

$$L_1' = (n_1/n_2)(S_2/S_1)^{1/2} L_1 \tag{7-24}$$

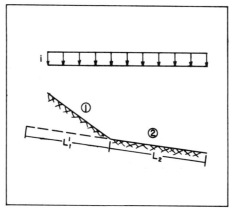

FIG. 7-13. A hillslope approximated as a cascade of two planes.

We now have an equivalent uniform plane, and the lagtime can be calculated in seconds as

$$t_L(\text{hillslope}) = (5/8q^{0.4})[(n_2/\sqrt{S_2})(L_1' + L_2)]^{0.6} \qquad (7\text{-}25)$$

Combining Eq. (7-24) and Eq. (7-25) leads to

$$t_L = (5/8q^{0.4})\{(nL/\sqrt{S})_1 + (nL/\sqrt{S})_2\}^{0.6} \qquad (7\text{-}26)$$

Variable Response Model

Convolution, routing, and hillslope transformation will now be combined to form the variable response model (VRM).

Consider the hillslope shown in Fig. 7-14. It can be partitioned into P-sections with

lengths	slopes	roughness
$l_1, l_2, l_3, \ldots, l_p$;	$S_1, S_2, S_3, \ldots, S_p$;	$n_1, n_2, n_3, \ldots, n_p$

Using the principles developed for cascading planes, the entire hillslope can be transformed into a uniform plane with roughness and dimensions of the lowest plane. For a steady uniform rain intensity, section 1 can be transformed into an equivalent length l_1' with the characteristics of section 2.

$$l_1' = l_1(S_2/S_1)^{1/2}(n_1/n_2) \qquad (7\text{-}27)$$

Total equivalent length of sections 1 and 2 is

$$l'_{12} = l_1' + l_2 \qquad (7\text{-}28)$$

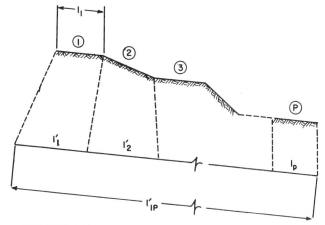

FIG. 7-14. A hillslope approximated as a cascade of P-planes.

Now, l'_{12} can be transformed into an equivalent length in terms of section 3.

$$l'_{13} = (l_1' + l_2)(n_2/n_3)(S_3/S_2)^{1/2} + l_3 \tag{7-29}$$

When this process is carried forward to section P, the total equivalent length of the entire hillslope in terms of section P is

$$l'_{1P} = l_P + \left(\frac{S_P^{1/2}}{n_P}\right) \sum_{x=1}^{P-1} \left(\frac{nl}{S^{1/2}}\right)_x \tag{7-30}$$

Lagtime for the equivalent plane for a steady uniform rain rate is

$$t_L = \frac{0.625(n_2 l'_{12})^{0.6}}{q^{0.4} S_P^{0.3}} \tag{7-31}$$

By substituting Eq. (7-30) into Eq. (7-31) and combining terms

$$t_L = \frac{0.625}{q^{0.4}} \left\{ \sum_{x=1}^{P} \left(\frac{nl}{S^{1/2}}\right)_x \right\}^{0.6} \tag{7-32}$$

Now, superimpose on this system of computations a steady nonuniform rainfall pattern:

$$q_1, q_2, q_3, \ldots, q_P$$

where q is the rain rate at a section in cubic feet per second per foot width per foot length of the section.

The total hydrograph at the end of the slope is calculated by

1. Transforming the hillslope into an equivalent plane for each section using Eq. (7-32);
2. Transmitting the hydrograph from each section to the end of the hillslope by the routing function Eq. (7-19); and
3. Summing up all transmitted hydrographs.

As an example, again consider the hillslope in Fig. 7-14 with a steady non-uniform rain pattern. Starting with section 1, lagtime for the equivalent plane is calculated using Eq. (7-32)

$$t_L = \frac{0.625}{q^{0.4}} \left\{ \sum_{x=1}^{P} \left(\frac{nl}{S^{1/2}}\right)_x \right\}^{0.6} \tag{7-33}$$

The routing function for section 1 is

$$R[l] = \left(\frac{l'_{1P}}{l_1}\right) Q_*[l'_{1P}] - \left(\frac{l'_{1P} - l_1'}{l_1'}\right) Q_*[l'_{1P} - l_1'] \tag{7-34}$$

where $Q_*[l'_{1P}]$ is the S-curve based on the lagtime for the total equivalent length, and $Q_*[l'_{1P} - l_1']$ is the S-curve based on the lagtime for the total equivalent length minus the equivalent length of the last section, l_1. This latter lagtime is calculated by summing from $x = 2$ to P in Eq. (7-32). The hydrograph transmitted from section 1 to the end of the plane is

$$Q[l/L] = q_1 l_1 R[l] \tag{7-35}$$

This procedure is continued for sections 2, 3, etc. The total hydrograph at the end of the plane becomes

$$Q[L] = \sum_{x=1}^{P} q_x l_x R[x] \tag{7-36}$$

where the routing function for each section is

$$R[x] = \left(\frac{l'_{xP}}{l_x'}\right) Q_*[l'_{xP}] - \left(\frac{l'_{xP} - l_x'}{l_x'}\right) Q_*[l'_{xP} - l_x'] \tag{7-37}$$

Combining Eqs. (7-37) and (7-36) results in

$$Q[L] = \sum_{x=1}^{P} q_x l_x \left[\left(\frac{l'_{xP}}{l_x'}\right) Q_*[l'_{xP}] - \left(\frac{l'_{xP} - l_x'}{l_x'}\right) Q_*[l'_{xP} - l_x']\right] \tag{7-38}$$

Lagtimes associated with the S-curves in Eq. (7-38) are calculated by Eq. (7-33) arranged in general terms as

$$t_L(l'_{xP}) = \frac{0.625}{q^{0.4}} \left\{\sum_{m=x}^{P} \left(\frac{nl}{S^{1/2}}\right)_m\right\}^{0.6} \tag{7-39}$$

and

$$t_L(l'_{xP} - l_x') = \frac{0.625}{q^{0.4}} \left\{\sum_{m=x+1}^{P} \left(\frac{nl}{S^{1/2}}\right)\right\}^{0.6} \tag{7-40}$$

Therefore, Eqs. (7-38)–(7-40) comprise the mathematical model for a steady, nonuniform rain pattern on a hillslope. Again, the S-curve is represented by Eq. (7-7) with $B = 0.30$.

The variable response model was completed by superimposing the model developed for unsteady rainfall on a uniform plane, i.e., by convolution, Eq. (7-16). The analysis just performed has theoretically transformed the nonuniform hillslope into a uniform plane. It only remains to convolute an applied unsteady rainstorm. Again, consider the hillslope in Fig. 7-14. An unsteady rain pattern is applied to each section tabulated at equal time intervals Δt.

$$q(1, j), \quad q(2, j), \quad q(3, j), \ldots, q(P, j)$$

where j represents the rain rate at the xth section during the jth rain period, or time interval. The rain rate in space and in time is represented in discrete form as $q(x, j)$.

The routing function for each section is now time dependent since rain rate is time dependent. Equation (7-34) becomes

$$R[x, j] = \left(\frac{l'_{xP}}{l_x}\right)Q_*[l'_{xP}, j] - \left(\frac{l'_{xP} - l'_x}{l'_x}\right)Q_*[l'_{xP} - l'_x, j] \qquad (7\text{-}41)$$

where $Q_*[l'_{xP}, j]$ refers to the S-curve for the equivalent plane, x to P, associated with a lagtime due to the rain rate on section x that occurred during the jth time interval. Hydrographs at each section must be synthesized, transmitted, and summed to form the total hydrograph. The transmitted hydrograph from section x to section P would be [using the principle of convolution, Eq. (7-19)]

$$Q[x/P, k\,\Delta t] = \sum_{j=1}^{k} q(x, j)l_x\{R[x, (k + 1 - j)\,\Delta t] - R[x, (k - j)\,\Delta t]\} \quad (7\text{-}42)$$

It is necessary to sum all transmitted hydrographs to obtain the total hydrograph at the end of the plane

$$Q[L, k\,\Delta t] = \sum_{x=1}^{P} \sum_{j=1}^{k} q(x, j)l_x\{R[x, (k + 1 - j)\,\Delta t] - R[x, (k - j)\,\Delta t]\}$$
$$(7\text{-}43)$$

By combining Eqs. (7-30) and (7-32) with Eq. (7-43), the final form of the variable response model can be used to compute discharge at the end of the plane at the kth time interval.

$$Q[L, k] = \sum_{x=1}^{P} \sum_{j=1}^{k} \lambda q(x, j)\{Q_*[l'_{xP}, k + 1 - j] - Q_*[l'_{xP}, k - j]$$

$$- \left(1 - \frac{l_x}{\lambda}\right)(Q_*[l'_{xP} - l'_x, k + 1 - j]$$

$$- Q_*[l'_{xP} - l'_x, k - j])\} \qquad (7\text{-}44)$$

where

$$\lambda = \frac{S_x^{1/2}}{n_x} \sum_{m=x}^{P} \left(\frac{nl}{S^{1/2}}\right)_m \qquad (7\text{-}45)$$

Equations (7-39) and (7-40) can still be used to calculate the appropriate lagtimes for the four S-curves in Eq. (7-44). However, it was more convenient to compute in minutes and operate with rain rates in inches per hour. Making these transformations, Eqs. (7-39) and (7-40) become

$$t_L[l'_{xP}, j] = \frac{0.58}{i[x, j]^{0.4}} \left\{ \sum_{m=x}^{P} \left(\frac{nl}{S^{1/2}} \right)_m \right\}^{0.6} \tag{7-46}$$

and

$$t_L[l'_{xP} - l'_{x'}, j] = \frac{0.58}{i[x, j]^{0.4}} \left\{ \sum_{m=x+1}^{P} \left(\frac{nl}{S^{1/2}} \right)_m \right\}^{0.6} \tag{7-47}$$

Equations (7-44)–(7-47) comprise the variable response model (VRM) of overland flow. Although at a glance the system of computations appears cumbersome, it is actually quite simple.

The key to the use of the VRM is the prediction of lagtime by Eqs. (7-46) and (7-47). Using the resistance coefficients determined from normal flow, predicted values of lagtime were correlated with the observed lagtimes taken from the 214 equilibrium hydrographs of the Corps of Engineers. The results are shown in Fig. 7-15. The correlation produced a 14% standard error. However, there is a deviation from the line of equal values at high values of lagtime. This bias could be explained by two effects. Observed lagtimes are consistently lower than predicted lagtimes. In determining observed lagtimes from equilibrium hydrographs, such as the schematic shown in Fig. 4-2, some error is introduced by the small portion of

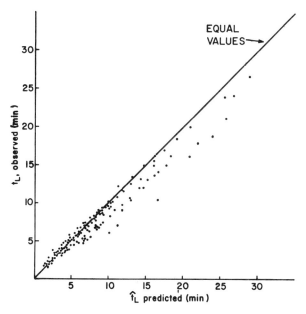

FIG. 7-15. Observed versus predicted lagtimes of 214 Corps of Engineers' data.

equilibrium storage present in area III; if the duration of rain had been longer, this error would have been diminished. Since lagtime is proportional to equilibrium storage, and since full equilibrium storage had not been reached, the lagtime determined would be less than the actual. A second effect could be attributable to a discrepancy between the Manning n-value of uniform flow and the n-value of equilibrium hydrographs. Also, it is possible that n-value significantly changes with depth in spatially varied flow.

The error is not considered to be serious in view of the other approximations made and the realization that the VRM will be extrapolated. Therefore, no correction was applied to Eqs. (7-46) and (7-47).

7-5   Examples of Application of VRM

Several examples have been selected for illustrating how the VRM can compute catchment hydrographs generated by complex rain patterns on complex geometries. The observed hydrographs were measured by the Corps of Engineers.

Uniform Unsteady Rainfall on a Uniform Plane

EXAMPLE 7.2   Computation of a recession of an equilibrium hydrograph on the 1% concrete through is shown in Fig. 7-16. The observed drops faster than the VRM. Such a fast drop of the observed could be

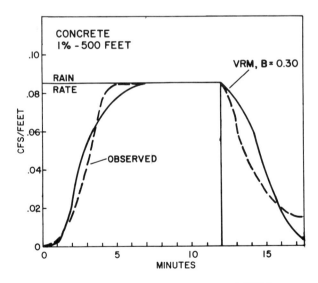

FIG. 7-16. Prediction of a recession with VRM.

explained in terms of kinematic wave theory in that it is nearly a translatory process. The computed recession, being just the inverse of the rising hydrograph, cannot drop suddenly. This initial drop of the VRM recession is rounded because initially there is a very slow rise on the rising hydrograph and through convolution (offsetting and subtracting S-curves) this effect is induced.

EXAMPLE 7.3   The effect just discussed also resulted in computation of partial run hydrographs. As shown in Fig. 7-17, the observed drops quickly whereas the peak of the VRM is more rounded.

EXAMPLE 7.4   A switch run hydrograph computed by the VRM is shown in Fig. 7-18. The recession effect mentioned above is present here. The VRM overshot the second equilibrium rate and undershot the third equilibrium rate. This effect can be reduced further by increasing the B-value, but some accuracy in computing the rising hydrograph would be sacrificed.

EXAMPLE 7.5   The recession effect and the undershooting effect resulted in the computation of a simulated storm shown in Fig. 7-19; but the VRM agrees closely with the observed. It appears that with a change in intensity the observed rises faster than it does on an initially dry plane. This could be because the state of flow is already fully turbulent when antecedent flows are present. The turbulent kinematic wave solution rises initially faster than the viscous kinematic wave solution.

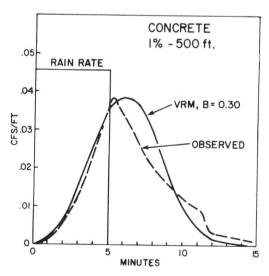

FIG. 7-17. Prediction of a partial run with VRM.

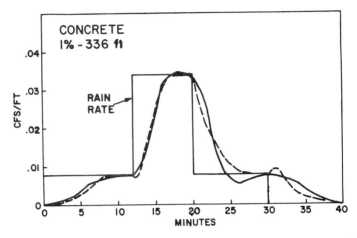

FIG. 7-18. Prediction of a switch run with VRM. --- observed; —— VRM; $B = 0.30$.

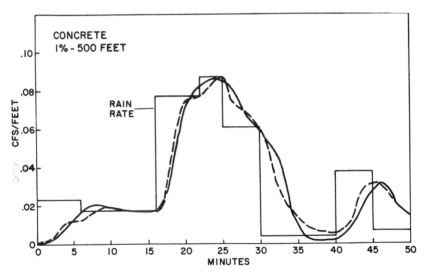

FIG. 7-19. Prediction of a simulated storm with VRM. --- observed; —— VRM; $B = 0.30$.

Steady Nonuniform Rainfall on a Uniform Plane

EXAMPLE 7.6 A hydrograph was measured at 168 ft which was generated by a steady rain intensity of 6.44 in./hr on the upper 84 ft only. Shown in Fig. 7-20 are the observed and VRM hydrographs at 168 ft. The VRM rises earlier but follows the recession closely. The discrepancy on the rise

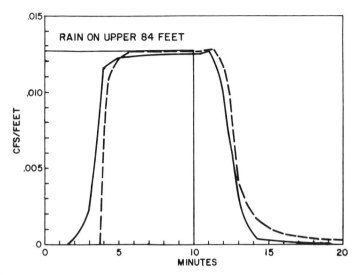

FIG. 7-20. Prediction of a hydrograph at 168 ft generated by rainfall over upper 84 ft with VRM. --- observed; —— VRM; $B = 0.30$.

is due to the lack of information of the hydraulic processes that occur when water flows over an initially dry channel.

Steady Uniform Rainfall on a Nonuniform Plane

EXAMPLE 7.6 A hydrograph was measured at 500 ft which was generated by a steady uniform rain rate of 3.96 in./hr. The surface of the upper 164 ft was concrete and the lower 336 ft was roughened with simulated turf. This nonuniform plane was transformed into an equivalent uniform plane and a hydrograph was computed by the VRM. Both the observed and computed hydrographs are shown in Fig. 7-21. There are large errors in prediction; not only is the recession effect present, but the VRM rises too early. It is possible that an actual kinematic shock was produced at the transition from concrete to simulated turf, thereby producing the steep rise in the observed. The VRM cannot reproduce a kinematic shock effect.

Prediction

EXAMPLE 7.7 Izzard [8] conducted a set of experiments on asphalt and reported that all flows were in the laminar state. An attempt was made to compute one of Izzard's hydrographs to demonstrate that the VRM can be applied to other data sets. The main problem encountered was estimation of a Manning n-value. Chow [9] reports an n-value of 0.013 for asphalt.

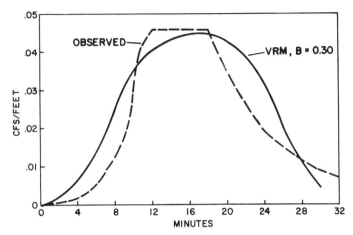

FIG. 7-21. Prediction of a hydrograph generated by a steady uniform rainfall on a non-uniform surface with VRM.

However, the lagtime predicted for the Izzard event was much lower than the observed; therefore, the observed and predicted lagtimes were set equal and an n-value of 0.024 was computed. This value was used in predicting the shape of the hydrograph shown in Fig. 7-22.

Simulation of Overland Flow on a Cascade of Planes—Comparison with Solution of Kibler–Woolhiser Model

EXAMPLE 7.8   Kibler and Woolhiser [10] presented an example using the shallow water equations. It involved the effect of a cascade slope

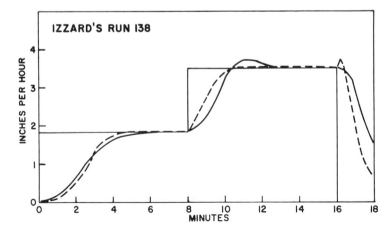

FIG. 7-22. Prediction of a switch run measured by Izzard with VRM. (After Izzard [8].) --- observed; —— VRM; $B = 0.30$; $L = 72$ ft; $S = 0.01$.

distribution on the overland flow hydrograph. They applied a steady uniform rain rate of 1 in./hr for 100 sec to five different three-plane cascades. They set the Chezy-C at 100 and the length of each plane at 33.33 ft, and computed the associated overland flow hydrographs using the second order Lax–Wendroff scheme. Their results are shown in Fig. 4-14. In general the shape of the rising hydrograph was similar to the shape of the hillslope. Kinematic shocks resulted on hillslopes 2, 3, and 4 and were smoothed out by the Lax–Wendroff scheme. An attempt was made to simulate this effect with the VRM.

In the VRM, lagtime has been placed in terms of a Manning n-value. An n-value corresponding to a Chezy-C of 100 was found by setting the time of equilibrium of the plane's rising hydrograph (hillslope 5) equal to the prediction equation of time to equilibrium in the VRM. This produced an n-value equal to 0.01. Hydrographs for the five cascades were computed by the VRM and are shown in Fig. 7-23. The VRM hydrographs are not as peaked as those in Fig. 4-14 but the shapes of these five hydrographs are in the same relative order as in Fig. 4-14. Shapes of the rising hydrograph from the VRM are similar to the shapes of the hillslopes.

Simulation of Moving Rainstorms on a Hillslope Formed as a Cascade of Planes

EXAMPLE 7.9 There are no data available to test the accuracy of the VRM except for the examples shown so far. To illustrate the flexibility of the VRM, the following example is presented. Consider the hillslope in

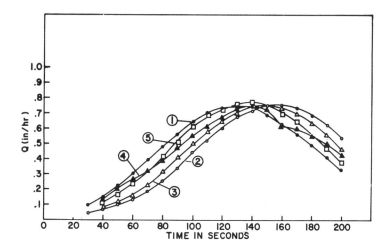

FIG. 7-23. Effect of slope shape on hydrographs using VRM. $B = 0.30$.

Fig. 7-24, and the associated rainfall hyetograph. Three examples will be computed:

1. an unsteady uniform storm as shown in the diagram,
2. the storm moving downslope offset by one minute, and
3. the storm moving upslope offset by one minute.

For storm 1, the rain will be uniform but unsteady as shown by the diagram in Fig. 7-24. For storm 2, rain will begin at $t = 0$ min at section 1, at $t = 1$ min at section 2, and $t = 2$ min at section 3. This was reversed for storm 3. The calculated hydrographs for these three storms are shown in Fig. 7-25. As the storm moves downslope, the wave builds up and a higher and later peak relative to the stationary storm should result. As the storm moves upslope, the wave cannot build up as high because the rain is moving away from the wave. The approximate speed of these storms was 3 miles/hr. Different results could be obtained for other hyetographs and storm speeds.

7-6 Hydrologic Design of Stormwater Inlets

Ragan *et al.* [11] have developed a dimensionless inlet hydrograph

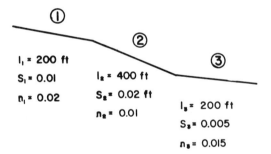

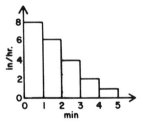

FIG. 7-24. A nonuniform hillslope with an unsteady nonuniform rainstorm.

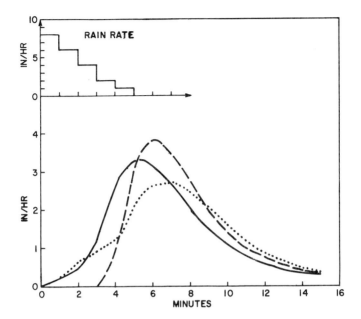

FIG. 7-25. Calculated hydrographs from hillslope in Fig. 7-24 using VRM. —— stationary; — ·— downslope (3 mph); . . . . upslope (3 mph); $B = 0.30$.

model (DIHM) which can be efficiently used in generating design storm-water flows into catch basins. The basic model links the VRM and Horton's infiltration model for calculating overland flow off rooftops, driveways, roads, and lawns with the kinematic equations for calculating flows in gutters; the gutter flow empties into the catch basin. This model was called the Maryland linked system design model (MLSDM). Jackson and Ragan [12] have also reported the use of the VRM in evaluating the hydrology of porous pavement parking lots.

The linkage of the VRM and Horton's model with the kinematic model for gutter flow was applied to inlet areas in Washington, D.C. for various design rainstorms. The results generated by MLSDM were then generalized and a dimensionless hydrograph was derived as a function of slope and percent imperviousness; this is the DIHM. The DIHM can then be used to generate a design hydrograph for any inlet for any design rainfall.

Favorable comparisons of the MLSDM with the EPA and University of Cincinnati models were made. The main advantage of DIHM is that design hydrographs based upon deterministic modeling can now be made efficiently using manual computations.

7-7 Conclusions

The purpose of the systems approach as applied to a deterministic treatment of stormwater is to develop eventually design or prediction methods which are rationally based and computationally sound, but which are reliable and easy to use. The ultimate goal of a systems type model is to link efficiently input (rainfall) to output (stormwater). This is effected by generalizing outputs developed from the deterministic models and then discarding the elaborate computational schemes which generated the output. The models developed, tested, and applied in this chapter were intended to illustrate the pragmatic nature of the systems approach to stormwater modeling.

Problems

7-1. Distinguish between linear and nonlinear system response both mathematically and physically.

7-2. Convolute the rain excess distribution

$$i = i_0(t/T) \qquad 0 \le t \le T$$
$$i = 0 \qquad\qquad t > T$$

with the instantaneous unit hydrograph

$$u = (1/T)e^{-t/T}$$

7-3. Attempt an explanation of the example of Minshall in terms of the deterministic model development and applications presented thus far.

7-4. Why was the IRF of the Manning-kinematic model not used in the VRM? What was substituted for it?

7-5. Derive the VRF shown in Fig. 7-9 using Fig. 7-8.

7-6. Derive Eq. (7-16).

7-8. Verify Figs. 7-11 and 7-12.

7-9. Conceptualize another derivation of the routing function R.

7-10. Apply the VRM to a parking lot in your area for a design rainfall event and compare the resulting stormwater hydrograph with the kinematic wave model.

7-11. Calculate the predicted hydrograph shown in Fig. 7-17 manually using the VRM.

7-12. What are some advantages and disadvantages of the VRM relative to other stormwater models presented in this text?

7-13. Account for the lower peak flows generated by the VRM in Fig. 7-23 relative to those generated by Kibler and Woolhiser in Fig. 4-14.

7-14. Further investigate the effect of rainfall spatial and temporal distributions on stormwater hydrographs generated on the hillslope in Fig. 7-24. Include the effects of moving rainstorms and generalize the results relative to the catchment characteristics.

7-15. Utilize the dimensionless inlet hydrograph method (DIHM) developed by Ragan *et al.* [11] to develop a design hydrograph for an inlet area in your town.

References

1. Corps of Engineers, U.S. Army, Data report, airfield drainage investigations, Los Angeles District, Office of the Chief of Engineers, Airfields Branch Engr. Div., Military Construction, Oct. 1954.

2. Sherman, L. K., The relation of hydrographs of runoff to size and character of drainage basins, *Trans. Amer. Geophys. Union*, 332–339 (1932).

3. Minshall, N. E., Predicting storm runoff on small experimental watersheds, *J. Hydr. Div., Amer. Soc. Civil Engr.* **86**, HY8, 17–38 (1960).

4. Overton, D. E., Analytical simulation of watershed hydrographs from rainfall, *Proc. Intern. Hydrol. Symp., Ft. Collins, Colorado*, 9–17 (Sept. 1967).

5. Overton, D. E., A variable response overland flow model, Ph.D. Dissertation, Dept. of Civil Engr., Univ. of Maryland, 1972.

6. Snyder, W. M., Extended continuous interpolation, *J. Hydr. Div., Amer. Soc. Civil Engrs.* **94**, HY5, 261–280 (1967).

7. Woolhiser, D. A., and Liggett, J. A., Unsteady one-dimensional flow over a plane—the rising hydrograph, *Water Resources Res.* **3**, No. 3, 753–771 (1967).

8. Izzard, C. F., Hydraulics of runoff from developed surfaces, *Proc. 26th Ann. Meeting Highway Res. Board, Washington, D.C.*, 129–150 (1946).

9. Chow, Ven Te, "Open Channel Hydraulics." McGraw–Hill, New York, 1959.

10. Kibler, D. F., and Woolhiser, D. A., The kinematic cascade as a hydrologic model, Hydrology Papers No. 39, Colorado State Univ., March 1970.

11. Ragan, R. M., Root, M. J., and Miller, J. F., Dimensionless inlet hydrograph model, *J. Hydr. Div., Amer. Soc. Civil Engr.*, **101**, HY9, 1185 (1975).

12. Jackson, T. J., and Ragan, R. M., Hydrology of porous pavement parking lots, *J. Hydr. Div., Amer. Soc. Civil Engr.*, **100**, HY12, 1739–1752 (1974).

PART III | Parametric Modeling

Chapter 8 | History of Parametric Stormwater Modeling

8-1 Introduction

Parametric modeling lies between deterministic and stochastic (purely random) modeling. In essence, the parametric approach strives for the definition of functional relations between hydrologic and geometric and land use characteristics of a catchment or watershed. This modeling approach involves the model formulation, data collection, data processing, model evaluation by optimization, regionalization of model parameters, and prediction of stormwater flows from ungauged watersheds.

The stochastic scale of modeling approaches is shown in Fig. 8-1. If the modeler has almost no information of cause–effect, then the stormwater process must be regarded as purely random. In this instance, the runoff process must be treated as being based entirely on chance. If the modeler has almost complete information on cause–effect, then the process may be treated as deterministic. Seldom, however, are we on the extremes of the scale, because we usually have some notion of the cause–effect of the process and seldom do we have the required information for a rigorous deterministic treatment of the process. The parametric approach is a compromise

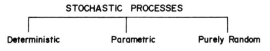

FIG. 8-1. The stochastic scale of modeling approaches.

between the two extremes, but it also involves an effort to improve our notions of the process. *Understanding of the process, in the opinion of these authors, is something we must take on faith.*

Another distinction among the modeling approaches can be made not only on the amount of, but also upon the type of information available. We must have boundary conditions and initial values in order to utilize the deterministic approach, but we do not need observed rainfall and runoff (hydrographs) data. In the parametric approach, it is imperative that we have observed rainfall and runoff data. In the stochastic (random) approach, we must have an observed time series of runoff, but it is not necessary that we have rainfall records [1]. However, in stochastic modeling, since there is no cause–effect concepts built into the model, analysis of the time series must be done only during periods of constant land use.

Figure 8-2 indicates the steps involved in parametric modeling; it must be emphasized that the process is *heuristic* and *iterative*. After parameter optimization, it may be concluded that the model has done a poor job of fitting the data, hence, adjustment of the model structure could be made and the experiment repeated. Further, it could be concluded that even though the model does a good job of fitting the data, little physical interpretation can be placed on the optimized model parameters. At this point, the model is a "black box" and any attempt to regionalize the parameters would be futile. Therefore, another adjustment of the model structure would be necessary and the experiment would be repeated. A sensitivity analysis would be included in the parameter interpretation.

Not all of the parametric stormwater models on the market utilize an objective best fit criteria in optimization; some have "eyeball" fits. The need for objective best fit criteria in parameter optimization is discussed in Chapter 9.

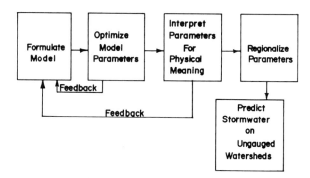

FIG. 8-2. Steps in parametric modeling.

8-2   Parametric Modeling Prior to the High Speed Digital Computer

Coefficient Methods-Peak Runoff

Perhaps the earliest attempt at parametric stormwater modeling was the use of coefficient methods for prediction of peak rates of stormwater runoff. The most widely used formula for peak runoff design is the rational formula [2] which has been previously discussed in Chapter 5. This method, like most coefficient methods predicts peak runoff rates by multiplying a coefficient with the design storm rainfall intensity. All losses are therefore lumped into the coefficient and no hydrograph shape is provided. Coefficient methods are inherently based upon linearity, and as shown in Chapter 5, the rational coefficient is rainfall dependent.

Unit Hydrograph Approach

Sherman [3] reported the concept of a unit hydrograph which was previously discussed in Chapter 7. The unit hydrograph concept merely states that for a given duration of rainfall and a constant land use and watershed condition, the same unit response will result for the basin. This concept is also based upon linearity, and as shown in the previous chapters, overland and channel response are nonlinear. The unit hydrograph approach is basically a "black box" model.

Overland Flow Hydrograph

Horton [4] improved the concepts of watershed runoff by postulating that there is a surface runoff or overland flow component, a channel flow component, and a groundwater return or base flow component. His model of watershed runoff was basically a kinematic flow approximation which has proven to be the forerunner of modern surface water hydraulics. Figure 8-3 illustrates Horton's concept of watershed runoff. He also postulated that overland flow ended on the inflection point on the recession, however, there was no evidence presented to substantiate the theory.

The Concept of Routing

Nash [5] developed an old concept of routing of effective rainfall through a series of linear reservoirs (Fig. 8-4). If the watershed flow system is conceptualized as a series of linear storage elements, the instantaneous unit hydrograph of the system is a gamma density function of the form

$$U(o, t) = \frac{1}{K\Gamma(N)}\left(\frac{t}{K}\right)^{N-1} e^{-t/K} \tag{8-1}$$

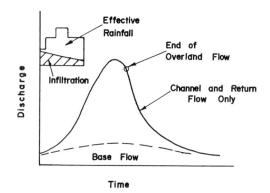

FIG. 8-3. Horton's concept of watershed runoff.

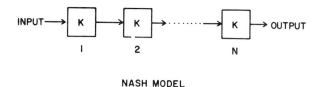

NASH MODEL

FIG. 8-4. The Nash model.

where N is the number of linear reservoirs and K the storage coefficient of each reservoir which relates the storage S in each element to its outflow Q.

$$S = KQ \qquad (8\text{-}2)$$

As shown in Fig. 8-5, the Nash model looks like a hydrograph. Equation (8-1) was plotted for comparison by holding the total lagtime of the system equal to

$$t_L = NK \qquad (8\text{-}3)$$

The Nash model has been a workhorse in watershed runoff computations. The principal question raised has been how many equal linear reservoirs are needed in using this model. Overton [6] reported an analysis of the Nash model and found that the model rapidly approached translation ($N \to \infty$) for a relatively small number of reservoirs. Most reported results indicate that from 1 to 5 reservoirs are satisfactory. An analysis of a very large number of storm hydrographs by Holtan and Overton [7] indicated that two reservoirs produced optimum results in fitting.

Maddaus and Eagleson [8] reported a model of equally distributed inputs into a series of equal linear reservoirs, as shown in Fig. 8-6. Holding

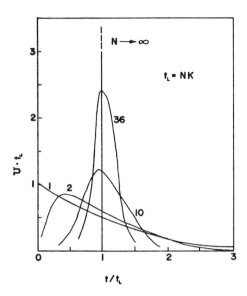

FIG. 8-5. Output from Nash model.

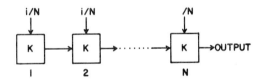

DISTRIBUTED INPUTS

FIG. 8-6. Linear model with equally distributed inputs.

lagtime constant for the system as was done with the Nash model, the instantaneous response function for the system was derived and is shown in Fig. 8-7. The response functions do not look like hydrographs and hence would be more of a "black box" than Nash's model. The response function for the linearly distributed model is

$$t_L U(o, t) = \frac{e^{-t/K}}{N} \sum_{j=1}^{N} \frac{1}{(N-j)!} \left(\frac{t}{K}\right)^{N-j} \tag{8-4}$$

In essence, these linear reservoir models are a special case of the kinematic wave model. A linear storage element would result if the exponent in Eq. (4-1) were set equal to unity. This move results in a tacit assumption that velocity is constant throughout the system and is equal to

$$V = a/w \tag{8-5}$$

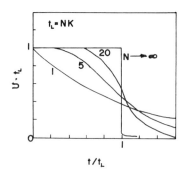

FIG. 8-7. Instantaneous response function of model shown in Fig. 8-6.

and storage in a linear element would be

$$S = ywL \qquad (8\text{-}6)$$

where L is the length of the system, and the reservoir coefficient is

$$K = wL/a \qquad (8\text{-}7)$$

Although assumption of linearity theoretically and experimentally has been shown to be incorrect, nevertheless, routing effective rainfall through a series of reservoirs has been shown to be a respectable approximation to stormwater. However, the parameters in the system will vary from storm to storm as demonstrated in Chapter 7.

8-3 Parametric Modeling with High Speed Digital Computers

It has become a matter of general agreement amongst water resource engineers that hydrologic information can be transferred from one basin to another only by mathematical modeling [10], but the very small amount of hydrologic data on small watersheds in the US has caused a serious limitation to development and testing of stormwater runoff models. It is most unlikely that any significant improvement in current models is possible until more and better quality data become available [11]. The key, therefore, to the evaluation of the effects of land use on stormwater runoff and quality is to analyze an adequate sample of hydrologic information to develop, test, and improve mathematical models of the rainfall–runoff quality regime of the basins studied.

In spite of the sparsity of hydrologic information, there are numerous models available for analysis of stormwater runoff [11]. Indeed, there are so many models available that some scientists and water resource engineers have called for a halt to development of new models and an intensification of testing, improving, and comparing existing models. This is a good principle; however, the structure of mathematical models depends upon the watershed to be modeled and the purpose of the study. Since no two watersheds or purposes are alike, it logically follows that no two mathematical models of watershed runoff should be alike. Hence, a trade off must be made between utilizing currently available modeling techniques and "tailoring" the model to the watershed and the purpose of the study.

Stanford Watershed Model

The Stanford Watershed Model [12] (SWM) was the earliest of the hydrologic models but has since been revised and improved several times. The most recent version is SWM IV [13] which has 21 parameters for rainfall and runoff simulation. SWM IV, as well as most existing models, was developed from an integration of component processes of the runoff phenomenon, i.e., infiltration, overland flow, groundwater, interflow, and channel routing. There are several modified versions of the SWM. The Kentucky Watershed model [14] and the Texas Watershed model [15] are examples of "tailoring" the SWM to particular needs in the Eastern and Southeastern United States. The SWM does not have procedures built in for optimizing the parameters by an objective best fit criteria, hence, much judgment must be employed to arrive at a "best" set of parameters. Recently, Monroe [16] utilized a pattern search procedure in optimizing the parameters in the National Weather Service version of SWM. Although the SWM is perhaps the most general model available, the main disadvantage in its use is the complexity involved, the large expense in computations, and the difficulty in examining any significant interrelationships amongst the parameters.

US Agricultural Research Service Model

The US Agricultural Research Service [17] reported a stormwater model which essentially computerized the (then) existing techniques utilized by the US Soil Conservation Service in their watershed flood control activities. The model utilized a unit hydrograph approach for subarea inputs and a storage technique for channel routing. The main emphasis of the model was evaluating the effects of potential reservoirs for peak runoff reduction. It was shown that it would not be possible to evaluate manually all of the alternatives evaluated by the computer.

US Geological Survey Model

The US Geological Survey model [18] is patterned after the SWM but is considerably less complex. It has only eight parameters but has the capability of simulating a continuous water balance. The model was developed primarily for the analysis of peak flows on rural areas and would require considerable modification if it were adapted for urban stormwater runoff use.

Purdue Model

Analysis of about two hundred storms in Indiana [19] using linear systems models indicated that the parameters in the models varied not only with an urbanization factor but also with other physiographic and meteorological factors. In particular, the model response parameters were found to be inversely related to the generating rain excess intensity. This finding closely corresponds with the well-known relation between time of concentration of overland flow and the generating rain intensity presented in previous chapters on deterministic models. This indicates that the system response could be approximated as linear within a storm but nonlinear from storm to storm. Yet, all components of the surface runoff process are known to be nonlinear. It has been shown that both lumped and distributed linear systems models are a poor approximation to overland flow [9]. In the linear models studied in Indiana, soil and other losses were estimated and were not optimized along with the surface response component. Hence, errors in soil loss estimation were absorbed into the optimized surface response parameters.

TVA Stormwater Model

A stormwater runoff model has been recently developed for evaluating the effects of rural land use on runoff in the Tennessee Valley [20]. The model is conceptually based upon the SCS soil loss and initial abstraction function presented in Chapter 2, and surface and ground water hydrograph response functions. The five model parameters are optimized by a pattern search procedure [21]. The model has been extensively tested on rural watersheds and a few small watersheds with a minor degree of urbanization. The optimized model parameters have been correlated with physiographic and meteorological factors.

TVA Daily Flow Model

A companion model to the TVA stormwater model is the TVA daily flow model [22]. The daily flow model serves as a stormwater model on larger basins as well as a continuous simulator of the moisture regime of a watershed. It has five parameters which are optimized by the pattern search method [21] and has been applied to a number of watersheds in the Tennessee Valley including a 382 sq mile basin which has been extensively stripped for surface coal.

There are many other parametric models. Several of these not mentioned here were elaborately compared by Linsley [11]. Few models have been extensively tested on hydrologic data. Because of wide variations in land use, climate and approach, comparisons of models are difficult and evaluations are usually subjective.

There does not seem to be a "perfect" model for analysis of stormwater. The models are either too complicated, do not allow for distributed inputs and parameters, do not simulate continuous streamflow, or have not been tested extensively on hydrologic data.

It is not feasible for the authors to expound upon all models reported in the literature. Instead, a few of the models have been selected for presentation in detail in order to demonstrate variations in modeling technique, application, and the type of inferences which may be drawn.

8-4   Present State of the Art

There remains much uncertainty in stormwater modeling. There does appear to be enough parametric models available which have been shown to be feasible conceptualizations of the stormwater runoff process. What is needed now is a continued and accelerated verification of the existing models and a follow-up regionalization of the parameters.

Problems

8-1.   Contrast parametric and deterministic stormwater modeling.

8-2.   What is a purely random process?

8-3.   Why is there a need for feedback in parametric stormwater modeling?

8-4.   What are the data requirements in parametric stormwater modeling?

8-5.   Derive Eqs. (8-1) and (8-4).

8-6.   Describe the Horton concept of stormwater. Can you defend it?

8-7.   Mathematically formulate your own parametric model for a real catchment or watershed.

References

1. Fiering, M. B., "Streamflow Synthesis." Harvard Univ. Press, Cambridge, Massachusetts, 1967.
2. Chow, V. T., "Handbook of Applied Hydrology." McGraw-Hill, New York, 1964.
3. Sherman, L. K., The relation of hydrographs of runoff to size and character of drainage basins, *Trans. Amer. Geophys. Union* 332–339 (1932).

4. Horton, R. E., The interpretation and application of runoff plot experiments with reference to soil erosion problems, *Soil Sci. Soc. Amer., Proc.* **3**, 340–349 (1938).

5. Nash, J. E., The form of the instantaneous unit hydrograph, *Proc. Gen. Assembly Toronto, Intern. Assoc. Scien. Hydrol.* **3**, 114–121 (1957).

6. Overton, D. E., Route or convolute? *Water Resources Res.* **6**, No. 1, 43–52 (Feb. 1970).

7. Holtan, H. N., and Overton, D. E., "Storage-flow hysteresis in hydrograph synthesis, *J. Hydrol.* **2**, 309–323 (1964).

8. Maddaus, W. O., and Eagleson, P. E., A distributed linear representation of surface runoff, Hydrodynamics Lab. Rept. No. 115, Dept. of Civil Eng., MIT, Cambridge, Massachusetts, June 1969.

9. Overton, D. E., A variable response overland flow model, Ph.D. Dissertation, Dept. of Civil Eng., Univ. of Maryland, 1971.

10. American Society of Civil Engineers, Urban Hydrology Research Council, "Basic Information Needs in Urban Hydrology," ASCE, New York, 1969.

11. Linsley, R. K., "A Critical Review of Currently Available Hydrologic Models for Analysis of Urban Stormwater Runoff," Hydrocomp International, August 1971.

12. Linsley, R. K., and Crawford, N. H., Computation of a synthetic streamflow record on a digital computer, *Intern. Assoc. Scien. Hydrol.* No. 51, 526–538 (1960).

13. Crawford, N. H., and Linsley, R. K., The synthesis of continuous streamflow hydrographs on a digital computer, Technical Report No. 39, Dept. of Civil Eng., Stanford University, 1966.

14. James, L. Douglas, An evaluation of relationships between streamflow patterns and watershed characteristics through the use of OPSET: A self calibrating version of the Stanford watershed model, Lexington, University of Kentucky, Water Resources Institute, Research Report No. 36, 1970.

15. Claborn, B. J., and Moore, W., Numerical simulation in watershed hydrology, Hydraulic Engineering Laboratory, University of Texas, Austin, Technical Rep. HYD 14-7001, 1970.

16. Monro, J. C., Direct search optimization in mathematical modeling and a watershed model application, NOAA Technological Memo NWS HYD-12, April 1970.

17. Holtan, H. N., and Overton, D. E., Numerical experiments in generating and routing watershed runoff, *Trans. Amer. Soc. Agr. Engr.,* 402–408 (1963).

18. Lichty, R. W., Dawdy, D. R., and Bergman, J. M., Rainfall-runoff model for small basin flood hydrograph simulation, *Proc., UNESCO–IASH Symp. Use Analog Digital Computers Hydrol.* **2** (1969).

19. Rao, R. A., Delleur, J. W., and Sparma, B. S. P., Conceptual hydrological models for urbanizing basins, *J. Hydr. Div., ASCE* **98** (HY7) 1205–1220 (July 1972).

20. Ardis, C. V., Jr., Storm hydrographs using a double triangle model, TVA, Division of Water Control Planning, Knoxville, Tennessee, Jan. 1973.

21. Green, R. F., Optimization by the pattern search method, TVA, Division of Water Control Planning, Knoxville, Tennessee, January 1970.

22. Betson, R. P., Upper Bear Creek experimental project-A continuous daily-streamflow model, TVA, Knoxville, Tennessee, Feb. 1972.

Chapter 9 | Model Optimization Techniques

9-1 Need for Objective Best Fitting

As discussed in Chapter 1, model parameter optimization is most efficiently and consistently achieved by objective best fitting. In this process, there are two fundamental decisions which must be made:

(1) what to use for an objective function, i.e., the criterion for measuring best fit, and

(2) what technique to use for achieving the optimum set of parameters that will satisfy the objective function.

Although there are numerous best fit criteria and numerous techniques for achieving the best fit, *it is important to understand that the model structure must be mathematically and computationally compatible with the objective best fit technique.* If this is not apparent to the reader at this point, then hopefully it will be after studying the examples in the chapter.

In this chapter, discussion will center upon several example objective best fit techniques which are widely used in stormwater modeling, and hence no attempt will be made to present a compendium of methodology.

9-2 Linear Least Squares

Perhaps the simplest best fit technique is linear least squares, but to be applicable the model optimized must be linear. This is the first illustration of the general principle stated above that the model must be compatible with the best fit technique.

The best fit criterion is to minimize the sum of the squares between the observed stormwater discharge Q and the fitted model discharge $\hat{Q}$. Mathematically, this is stated as

$$\text{minimize} \quad \left\{ S = \sum_{j=1}^{n} (Q - \hat{Q})_j{}^2 \right\} \qquad (9\text{-}1)$$

where n is the number of observations. Equation (9-1) is the general least squares best fit criterion.

Linear least squares is applicable to a linear model whereby stormwater discharge is conceptualized as a linear sum of m variates.

$$\hat{Q} = c_0 + c_1 x_1 + \cdots + c_m x_m \qquad (9\text{-}2)$$

where x_i are the independent hydrologic or physiographic variates which are being related to stormwater.

The objective function then becomes

$$S = \sum_{j=1}^{n} \left\{ Q - c_0 - \sum_{k=1}^{m} c_i x_i \right\}^2 \qquad (9\text{-}3)$$

The solution for the optimum set of coefficients is found by taking the first derivative of S with respect to each coefficient, setting them all to zero, and solving the resulting equations simultaneously for the coefficients c_i.

$$\frac{\partial S}{\partial c_0} = -2 \sum_{j=1}^{n} \left\{ Q - c_0 - \sum_{k=1}^{m} c_i x_i \right\} = 0 \qquad (9\text{-}4a)$$

and

$$\frac{\partial S}{\partial c_i} = -2 \sum_{j=1}^{n} \left\{ Q - c_0 - \sum_{k=1}^{m} c_i x_i \right\} \{x_i\} = 0 \qquad (9\text{-}4b)$$

Equations (9-4) are called the characteristic equations. They are linear and solution for the coefficient vector is readily obtained.

It should be obvious at this point that the linear least squares technique is only compatible with a linear model.

There are three basic assumptions involved in linear least squares:

(1) There is no error in the independent variates, hence, all errors are in the dependent variate.

(2) All independent variates are statistically uncorrelated.

(3) The errors in the dependent variates are normally distributed.

The measure of goodness of fit in the least squares function is the coefficient of determination r^2 which is the ratio of the variance explained by the model relative to the variance of the dependent variate.

$$r^2 = 1 - (SEE/SQ) \qquad (9\text{-}5)$$

where SEE is the error variance,

$$SEE = \frac{1}{n} \sum_{j=1}^{n} (Q - \hat{Q})^2 \qquad (9\text{-}6)$$

and SQ is the variance of the dependent variate, stormwater,

$$SQ = \frac{1}{n} \sum_{j=1}^{n} (Q - \bar{Q})^2 \qquad (9\text{-}7)$$

where $\bar{Q}$ is the mean of the observations of the dependent variate. The coefficient of determination varies from zero to unity and signifies no correlation and a perfect correlation, respectively.

EXAMPLE 9.1   For the monthly rainfall and runoff volumes given, use linear least squares to evaluate the model

$$\hat{Q}(3) = c_1 P(1) + c_2 P(2) + c_3 P(3)$$

The data for the White Hollow Watershed is shown in Table 9-1. The objective of the model is to evaluate December runoff volume as a function of December, November, and October rainfall.

The characteristic equations for the model are

$$\sum Q P(1) - c_1 \sum P(1)^2 - c_2 \sum P(1)P(2) - c_3 \sum P(1)P(3) = 0$$

$$\sum Q P(2) - c_1 \sum P(1)P(2) - c_2 \sum P(2)^2 - c_3 \sum P(2)P(3) = 0$$

$$\sum Q P(3) - c_1 \sum P(1)P(3) - c_2 \sum P(2)P(3) - c_3 \sum P(3)^2 = 0$$

TABLE 9-1

Year	Rainfall (in.)			Runoff (in.) December
	October	November	December	
1935	1.31	7.35	2.17	0.76
1936	2.61	2.42	8.17	1.21
1937	0.08	1.56	3.27	1.05
1938	0.16	4.98	2.62	0.57
1939	0.86	1.49	2.90	0.24
1940	2.01	2.29	2.55	0.41

In matrix form, the equations become

$$\begin{pmatrix} 13.34 & 22.75 & 32.47 \\ 22.75 & 94.58 & 64.03 \\ 32.47 & 64.03 & 103.93 \end{pmatrix} \begin{pmatrix} c_1 \\ c_2 \\ c_3 \end{pmatrix} = \begin{pmatrix} 5.36 \\ 14.29 \\ 18.20 \end{pmatrix}$$

and the solution is

$$\hat{Q}(3) = -0.16\,P(1) + 0.065\,P(2) + 0.19\,P(3)$$

The model shows some logic in that the weighting coefficients are decreasing in value. However, the negative coefficient on October rainfall seems illogical. There are at least three basic reasons for these results. First, the process is not linear; second, the data base is small and hence heavily biased; and, third, a rainfall persistence effect in the Fall months would tend to produce some interrelations amongst the independent variates thereby violating the assumptions of the model. This latter point will be addressed in the section on components analysis.

The standard error of estimates SEE is equal to 0.04, the standard deviation of discharge SQ is 0.116, and the coefficient of determination r^2 is 0.65. This means that the model has explained 65% of the variance of the December runoff.

9-3   Nonlinear Least Squares

Due to the limitations imposed by linearity, techniques for optimizing nonlinear models have been developed. "The Method of Differential Correction" [1] is a method which has been used considerably in hydrologic studies and was computerized and tested by DeCoursey and Snyder [2].

If the nonlinear model is expressed as

$$\hat{Q} = f(x, a) \tag{9-8}$$

where x is an independent variate and a a coefficient relating $\hat{Q}$ with x, the objective of the procedure is to find a better value of a from an initial estimate of the parameter. The relationship between Q and a is shown in Fig. 9-1.

Given an initial estimate of a, a_e, the estimated value of Q is $\hat{Q}$. The adjustment in a_e, necessary to calculate Q_0 is given as

$$Q_0 = \hat{Q} + h(dQ/da) + \delta_0 \tag{9-9}$$

where δ_0 is an error due to the curvature of the function. If the error is small relative to $h(dQ/da)$, then

$$Q_0 = \hat{Q} + h(dQ/da) \tag{9-10}$$

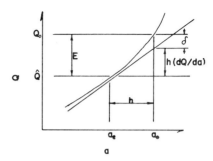

FIG. 9-1. Relationship between Q and a.

For a model with m parameters,

$$\hat{Q} = f(x_1, x_2, \ldots, x_m, a_1, a_2, \ldots, a_m) \tag{9-11}$$

and where the independent variates have observed field values, it follows that

$$Q_0 = g(a_1 + h_1, a_2 + h_2, \ldots, a_m + h_m) \tag{9-12}$$

The model, Eq. (9-12), can be expanded as a Taylor series as

$$Q_0 = \sum_{k=0}^{N-1} \frac{1}{k!} \left(h_1 \frac{\partial}{\partial a_1} + \cdots + h_m \frac{\partial}{\partial a_m} \right)^k \times g + R_N \tag{9-13}$$

where R_N is the remainder term.

When using an iterative technique in which h_j is used to adjust a_j, the value used as the new estimate, the remainder term may be ignored. Also, the first term of the model in Eq. (9-13) is $\hat{Q}$. Hence, letting E be equal to $Q_0 - \hat{Q}$, it follows that

$$E = h_1(\partial\hat{Q}/\partial a_1) + \cdots + h_m(\partial\hat{Q}/\partial h_m) \tag{9-14}$$

In situations where the model would not be in a continuous mathematical form, the error term can be utilized in finite form as

$$E = h_1(\Delta\hat{Q}/\Delta a_1) + \cdots + h_m(\Delta\hat{Q}/\Delta a_m) \tag{9-15}$$

where $\Delta\hat{Q}$ is the difference between the predicted value of Q, and the predicted value of Q with a_j incremented a small amount.

The nonlinear least squares objective function is

$$\text{minimize} \left\{ \sum_{j=1}^{n} E_j^2 \right\} \tag{9-16}$$

where n is the number of observations, and since the model has now been transformed into a linear function, i.e., Eq. (9-15), it follows that a nonlinear model may be optimized with linear least squares.

The optimization technique proceeds by utilizing a multiple regression in solving for h_j by the objective function, Eq. (9-16). The values of h_j are then added to a_i and the new values of a_i are used as a better predictor of Q_i. This iterative process is repeated until the algorithm converges and yields optimized parameter values by satisfying the objective function. To trigger the solution, an initial guess as to the "true" values of the parameters must be made. Further, the divided differences in the model, Eq. (9-15), may be considered as a measure of the sensitivity of the function with respect to the a_{ith} parameter.

The limitations imposed by the assumptions inherent to linear least squares also apply to the nonlinear least squares. The most important and limiting assumption is that the variates are statistically unrelated. If the terms in Eq. (9-15) are statistically interrelated, convergence of the method cannot be assured; but, even if convergence were assured, the method would tend to distribute the correlation of two interrelated independent terms with the dependent variate over the two independent terms. Hence, the resulting optimized parameters would look strange and little hope would exist for attributing physical significance to these parameters. And, even with an excellent model fit, the best that could be hoped for would be a "black box" model.

9-4   Principal Components Analysis

A very powerful technique has been utilized for coping with the problems in both linear and nonlinear least squares associated with statistical inter-relations amongst the independent variates. The technique, principal components analysis, transforms the independent variates in Eq. (9-2) for the linear models and in Eq. (9-15) for nonlinear models into new variates called "components" which are linear sums of the original variates. A search technique is employed to locate the components such that they are not statistically correlated. This provides us with new variates to correlate with stormwater Q which are truly independent, i.e., statistically unrelated. When the correlation with Q is completed, the components can be transformed back to the original variates.

Let us work with the linear model

$$\hat{Q} = c_1 x_1 + \cdots + c_P x_P \qquad (9\text{-}17)$$

and transform it by components analysis to

$$\hat{Q} = \beta_1 \zeta_1 + \cdots + \beta_P \zeta_P \qquad (9\text{-}18)$$

where ζ_i is a component or eigenvector. The components are new variates which are linear functions of x_i (the original variates) and they are statistically independent. Statistical independence is defined as having a covariance (cov) equal to zero.

$$\mathrm{cov}\{\zeta_\alpha \zeta_\beta\} = \sum_{j=1}^{n} \zeta_{\alpha j}\zeta_{\beta j} = 0 \qquad (9\text{-}19)$$

Our problem is made simpler by removing the scale effects of the original variates. Hence, the normalized original variates are defined as

$$\psi_i = (x_i - \bar{x})/s_i \qquad (9\text{-}20)$$

letting the mean be zero and the standard deviation s be unity. The first two moments of the normalized variates are

$$\sum_{j=1}^{n} \psi_{ij} = 0 \quad (\text{mean}) \qquad (9\text{-}21)$$

and

$$\frac{1}{n}\sum_{j=1}^{n} \psi_{ij}^2 = 1 \quad (\text{standard deviation}) \qquad (9\text{-}22)$$

The solution to our problem begins by "plotting" all p of the original variates in p-dimensional space and rotating the axes until the orthogonal system of components are found. An attempt to demonstrate this for three dimensions is shown in Fig. 9-2. The data points are plotted as referenced to the axes of the three original variates and then the axes are rotated until

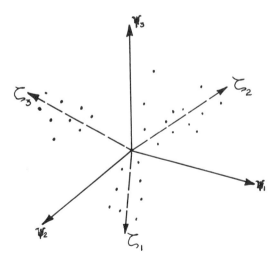

FIG. 9-2. Location of components in three dimensions.

the components are orthogonal, i.e., statistically independent. Statistically, this feat is achieved by minimizing the variance or spread around the components subject to the constraint that orthogonality must be achieved.

The relations between the normalized original variates and the components are demonstrated in Fig. 9-3 considering a single observation (ψ_{11}, ψ_{21}). The location of the component will be achieved by minimizing the variance ΔS^2. From geometry,

$$\Delta S^2 = AD^2 - AC^2 \tag{9-23}$$

and

$$\Delta S^2 = (\psi_{11}^2 + \psi_{21}^2) - (\psi_{11} \cos \theta_1 + \psi_{21} \cos \theta_2)^2 \tag{9-24}$$

Recognizing that

$$\cos^2 \theta_1 + \cos^2 \theta_2 = 1 \tag{9-25}$$

and letting l_1 and l_2 represent the cosine terms, Eq. (9-24) can be generalized for p-variates as

$$\Delta S^2 = \sum_{i=1}^{p} \psi_{i1}^2 - \left\{ \sum_{i=1}^{p} \psi_{21} l_i \right\}^2 \tag{9-26}$$

and for n-observations, Eq. (9-26) becomes

$$n \cdot \Delta S^2 = \sum_{j=1}^{n} \left\{ \sum_{i=1}^{p} \psi_{ij}^2 - \left\{ \sum_{i=1}^{p} l_i \psi_{ij} \right\}^2 \right\} \tag{9-27}$$

Recalling Eq. (9-22) and rearranging,

$$p - \Delta S^2 = \frac{1}{n} \sum_{j=1}^{n} \left\{ \sum_{i=1}^{p} l_i \psi_{ij} \right\}^2 \tag{9-28}$$

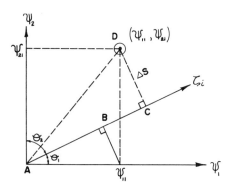

FIG. 9-3. Relation between normalized original variates and components.

Thus, $p - \Delta S^2$ is a variance of the variate on the right-hand side of Eq. (9-28) which is a component ζ_i.

Using only two variates, Fig. 9-4 can be used to demonstrate the relations between the original normalized variates and the component.

EXAMPLE 9.2 What is the degree of association of the component in Fig. 9-4 with the original normalized variates?

For the angles shown, $l_1 = 0.984$ and $l_2 = 0.173$ and this shows a close association of ζ_i with ψ_1 but not ψ_2. Geometrically, ζ_i is made up of AB plus BC and it can be seen that AB is much greater than BC.

We now seek location of the components by rotating the axes subject to the axes being mutually orthogonal. Then our objective function is

$$\text{minimize} \left\{ \frac{1}{n} \sum_{j=1}^{n} \left\{ \sum_{i=1}^{p} l_i \psi_{ij} \right\}^2 - \lambda \left\{ \sum_{i=1}^{p} l_i^2 - 1 \right\} \right\} \tag{9-29}$$

where λ is a Lagrangian multiplier or an eigenvalue. The second term forms the constraint of orthogonality.

The axes are efficiently rotated to achieve solution of Eq. (9-29) using a method called VARIMAX ROTATION.

Where V is the function minimized in Eq. (9-29), solution of the directional cosines is found by

$$\frac{\partial V}{\partial l_k} = -2\lambda l_k + \frac{2}{n} \sum_{j=1}^{n} \sum_{i=1}^{P} l_i \psi_{ij} \psi_{kj} = 0 \tag{9-30}$$

which can be rearranged as

$$-\lambda l_k + \frac{1}{n} \sum_{i=1}^{P} l_i \sum_{j=1}^{n} \psi_{ij} \psi_{kj} = 0 \tag{9-31}$$

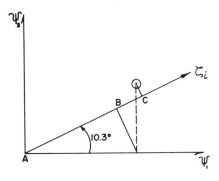

FIG. 9-4. Degree of association of the original normalized variates and the component.

We recognize the simple correlation coefficient, r_{ik} as

$$r_{ik} = \sum_{j=1}^{n} \psi_{ij}\psi_{kj} \tag{9-32}$$

Then Eq. (9-31) becomes

$$-\lambda l_k = \frac{1}{n} \sum_{i=1}^{P} l_i r_{ik} = 0 \tag{9-33}$$

Equation (9-33) is but a form of the simple correlation matrix with λ in the diagonal. Equation (9-33) can be expanded and rearranged as

$$\begin{pmatrix} (1-\lambda) & r_{12} & r_{13} & \cdots & r_{1P} \\ r_{21} & (1-\lambda) & r_{23} & \cdots & r_{2P} \\ r_{31} & r_{32} & (1-\lambda) & \cdots & r_{3P} \\ \vdots & & & & \\ r_{P1} & & \cdots & & (1-\lambda) \end{pmatrix} \begin{pmatrix} l_1 \\ l_2 \\ l_3 \\ \vdots \\ l_P \end{pmatrix} = 0 \tag{9-34}$$

Equation (9-34) can be written as

$$|r - \lambda I|\mathbf{l} = 0 \tag{9-35}$$

Since the directional cosine vector is nonzero, it follows that

$$|r - \lambda I| = 0 \tag{9-36}$$

Solution to the problem is by way of the eigenvector–eigenvalue route. There are p-roots to Eq. (9-36) and they are found by expanding Eq. (9-36) as a determinant and obtaining an algebraic equation to the pth power.

$$c_1\lambda^p + c_2\lambda^{p-1} + \cdots + c_{p+1} = 0 \tag{9-37}$$

Principal components analysis forces out the root that has the largest values first. Since λ is a variance of a component, forcing out the largest values of λ first means that the important components come out first, and hence this provides a mechanism for screening out marginal information.

At this point we should check for orthogonality of the components. Beginning with Eq. (9-33) and Fig. 9-4,

$$\lambda_\alpha l_{\alpha i} = \sum_{i=1}^{P} l_{\alpha i} r_{ij} \tag{9-38}$$

and

$$\lambda_\beta l_{\beta i} = \sum_{i=1}^{P} l_{\beta i} r_{ij} \tag{9-39}$$

multiply Eq. (9-38) by $l_{\beta i}$ and Eq. (9-39) by $l_{\alpha i}$. Then we sum each over the

n-observations and then subtract the two equations. This produces

$$(\lambda_\alpha - \lambda_\beta) \sum_{i=1}^{p} l_{\alpha i} l_{\beta i} = 0 \qquad (9\text{-}40)$$

Since the eigenvalues cannot be equal, it follows that

$$\sum_{i=1}^{p} l_{\alpha i} l_{\beta i} = 0 \qquad (9\text{-}41)$$

which therefore proves orthogonality. Further, since the components are orthogonal

$$\zeta_\alpha = \sum_{i=1}^{p} l_{\alpha i} \psi_\beta \qquad (9\text{-}42)$$

and

$$\psi_\alpha = \sum_{i=1}^{p} l_{i\beta} \zeta_\beta \qquad (9\text{-}43)$$

the covariance of the components is

$$\mathrm{cov}\{\zeta_\alpha \zeta_\beta\} = \mathrm{cov}\left\{ \sum_{i=1}^{p} l_{i\alpha} \psi_\alpha \sum_{i=1}^{p} l_{i\beta} \psi_\beta \right\} = 0 \qquad (9\text{-}44)$$

With the derivation of orthogonal components insured, we now need to concentrate on the principles of data interpretations whereby correlations of the original normalized variates with the components can be identified.

The variance of ψ_α is

$$V\{\psi_\alpha\} = V\left\{ \sum l_{i\beta} \zeta_\alpha \right\} \qquad (9\text{-}45)$$

where

$$V\{\psi_\alpha\} = \frac{1}{n} \{l_{11}\zeta_1 + l_{21}\zeta_2 + \cdots + l_{p1}\zeta_p\}^2 \qquad (9\text{-}46)$$

This results in

$$V\{\psi_\alpha\} = \{l_{11}^2 + l_{21}^2 + \cdots + l_{p1}^2\} \qquad (9\text{-}47)$$

or

$$V\{\psi_\alpha\} = \sum_{i=1}^{p} l_{i\alpha}^2 \qquad (9\text{-}48)$$

Hence, $l_{i\alpha}$ is a type correlation coefficient representing the correlation of ψ_β with ζ_α. The directional cosine squared is a variance, which represents the fraction of ψ_β explained by ζ_α.

Once principal components analysis has been completed, the regression can be performed whereby ζ_α is considered to be the independent variates. Then the model coefficient c_i in Eq. (9-17) can be derived from the regression coefficients in Eq. (9-18). Equating the two models results in

$$\begin{pmatrix} l_{11} & l_{21} & l_{p1} \\ l_{12} & l_{22} & \cdots & l_{p2} \\ \vdots & \vdots & \vdots \\ l_{1p} & l_{2p} & \cdots & l_{pp} \end{pmatrix} \begin{pmatrix} c_1 \\ c_2 \\ \vdots \\ c_p \end{pmatrix} = \begin{pmatrix} \beta_1 \\ \beta_2 \\ \vdots \\ \beta_p \end{pmatrix} \tag{9-49}$$

and the solution of c_i is

$$\mathbf{c} = L^{-1}\boldsymbol{\beta} \tag{9-50}$$

when components analysis is combined with nonlinear least squares, the original variates are $\partial\hat{Q}/\partial a_i$ and the coefficients are h_i.

A few examples will be shown which are intended to demonstrate the utility of components analysis.

EXAMPLE 9.2   This example was reported by Snyder [4] and involved the fitting of two equations to the data in Table 9-2, the only difference being the inclusion of the cross product term in Eq. (9-51). The equation used to predict the December runoff was

$$\hat{Q} = c_1 x_1 + c_2 x_2 + c_3 x_3 + c_4 x_4 + c_0 \tag{9-51}$$

where $\hat{Q}$ is the December runoff in inches, x_1 the December rainfall in inches, x_2 the November rainfall in inches, and x_3 the October rainfall in inches.

$$x_4 = x_1 \left(\frac{r_{Qx_2}x_2 + r_{Qx_3}x_3}{r_{Qx_2}} \right) \tag{9-52}$$

where r is a simple correlation coefficient.

There is no particular justification for the form of the model since the emphasis here is to compare the results of two methods of fitting. The results of fitting are shown in Table 9-3. When x_4 was added to the model, the results were no longer satisfactory. The first three parameters became negative, and only c_4 was positive. Transformation to normalized variates was no help because c_4 became large compared to the largest negative parameter. Intuitively, the contribution of the separate monthly rainfall terms should be positive even though the effect of each month's rainfall is included in the cross-product term.

The poor results of the regression when x_4 was included in the model were due to the high correlation of the term with rainfall of the separate

TABLE 9-2 Monthly rainfall and runoff of White Hollow watershed used in Example 9-2

| Year | Rainfall (in.) | | | | Runoff (in.) December |
	October	November	December	Weighted product	
1935	1.31	7.35	2.17	18.31	0.76
1936	2.61	2.42	8.17	32.00	1.21
1937	0.08	1.56	3.27	17.04	1.05
1938	0.16	4.98	2.62	13.31	0.57
1939	0.86	1.49	2.90	5.31	0.24
1940	2.01	2.29	2.55	8.92	0.41
1941	3.01	2.62	2.60	11.52	0.48
1942	1.77	1.79	8.68	24.74	2.62
1943	3.18	1.77	2.21	8.13	0.54
1944	0.92	3.78	3.86	16.71	1.49
1945	2.32	4.83	4.31	26.81	1.56
1946	4.19	3.60	3.25	19.39	0.86
1947	1.16	3.11	2.27	8.85	0.47
1948	1.32	10.02	5.07	54.91	1.82
1949	4.99	1.80	4.29	20.55	0.04
1950	1.07	3.03	3.37	12.37	1.03
1951	3.85	7.52	8.38	82.38	6.06
1952	1.04	5.11	4.10	23.49	0.65
1953	0.54	1.78	3.88	8.15	0.41
1954	2.16	2.38	8.66	36.20	1.39
1955	1.96	2.91	3.41	13.95	0.55
1956	1.84	2.43	6.25	23.00	1.13
1957	4.12	6.88	6.19	57.99	3.99
1958	1.50	3.01	2.98	11.65	0.43
1959	4.72	4.18	5.50	28.56	1.30

months. The interrelations built into the model has produced a "black box" and there is little hope of interpreting the optimized parameters. The solution by components regression with the x_4 term present has greatly improved the model.

In components regression, the end result desired is not simply to attain the highest possible model calibration. What is desired is a solution wherein the coefficients are sensible "weighting factors" for each term. Hence, the structure of the model is deemed more important than prediction with minimum error. This would be the case if the overall objective of the modeler would be to regionalize the model. In this situation, the model could be used as a reliable predictor on ungauged watersheds if physical significance could be attributed to the model parameters.

TABLE 9-3 Comparison of fittings of Eq. (9-51) in Example 9-2

Coefficient	Without x_4		With x_4	
	Original variates	Standardized variates	Original variates	Standardized variates
Part A. Multiple regression				
c_0	-1.688		0.942	
c_1	0.3384	0.5612	-0.1623	-0.2691
c_2	0.2634	0.4521	-0.2803	-0.4811
c_3	0.2058	0.2464	-0.1452	-0.1739
c_4			0.1030	$+1.4731$
Part B. Multivariate solution				
c_0	-1.690		-1.225	
c_1	0.3385	0.5613	0.2330	0.3559
c_2	0.2635	0.4522	0.1372	0.2254
c_3	0.2060	0.2467	0.1236	0.1433
c_4			0.0282	0.3967

EXAMPLE 9.3   Wong [5] developed a model for predicting mean annual floods in New England. The model was of the form

$$\hat{Q}_m = a x_1^b \, x_2^c \cdots x_{11}^l \qquad (9\text{-}53)$$

where $\hat{Q}_m$ is mean annual flood and x_1 through x_{11} were physiographic and rainfall variates. Wong transformed the model to a linear form by taking the natural logarithms of Eq. (9-53). This placed the model in the form of Eq. (9-51). He had predetermined Q_m and x_i for 90 watersheds in New England.

The 11 variates were (1) drainage area, (2) basin shape, (3) main channel slope, (4) tributary channel slope, (5) % of area in ponds and lakes, (6) average land slope, (7) mean altitude, (8) length of longest watercourse, (9) length of the main stream, (10) stream density, and (11) precipitation-frequency intensity. The simple correlation matrix of the 12 variates is shown in Table 9-4. There are several significant correlations of x_i with Q_m but at a glance it would be nearly impossible to decide which ones to choose. Further, there are a number of significant interrelations amongst the independent variates, but since the entire correlation procedure is a matter of degree, it would be impossible to filter out objectively the variates at this point. The overall correlation of the regressive model with the data was 0.97. Even though the model explained 94% of the variance of Q_m, it resulted in a "black box" model which severely limited its capability as a predictor on a regional basis. Further, even if the model could be reliably applied

TABLE 9-4 Simple correlation matrix of Example 9-3

Variates[a]	x_1	x_2	x_3	x_4	x_5	x_6	x_7	x_8	x_9	x_{10}	x_{11}	x_{12}
x_1	1	0.938	-0.464	-0.221	0.030	0.138	0.106	-0.309	0.913	0.926	-0.214	0.799
x_2		1	-0.539	-0.296	0.095	0.070	0.035	-0.381	0.961	0.958	-0.146	0.770
x_3			1	0.687	-0.506	0.511	0.536	0.538	-0.575	-0.539	-0.094	0.036
x_4				1	-0.596	0.866	0.809	0.509	-0.310	-0.300	-0.490	0.217
x_5					1	-0.628	-0.457	-0.216	0.141	0.159	0.294	-0.294
x_6						1	0.869	0.373	0.034	0.046	-0.604	0.517
x_7							1	0.410	0.018	0.097	-0.591	0.445
x_8								1	-0.347	-0.361	-0.013	-0.017
x_9									1	0.950	-0.145	0.719
x_{10}										1	-0.184	0.743
x_{11}											1	-0.345
x_{12}												1

[a] The variates are: x_1, drainage area; x_2, basin shape; x_3, main channel slope; x_4, tributary channel slope; x_5, % of area in ponds and lakes; x_6, average land slope; x_7, mean altitude; x_8, stream density; x_9, length of longest water course; x_{10}, length of the main stream; x_{11}, precipitation-frequency intensity; x_{12}, mean annual flood.

to an ungauged watershed, it would be very expensive to determine the values of the independent variates.

The resulting eigenvalues and VARIMAX normalized column eigenvectors are shown in Table 9-5. The numbers listed in the columns are the directional cosines which linearly link x_i with ζ_i. The eigenvalues are at the bottom of Table 9-5. Since

$$p = \sum_{i=1}^{p} \lambda_i \tag{9-54}$$

it follows that only four components account for $(10.9/12) \times 100\%$ or 90% of the explained variance of Q_m. Remembering that directional cosines are much like correlation coefficients, we see that component 1 is highly correlated with x_1, x_2, x_9, and x_{10} indicating that these four variates are highly interrelated. In component 2, we see that it is highly correlated with x_6 and x_7 indicating that these two variates are highly interrelated.

We can examine the variance explained by the components by squaring Table 9-5 shown in Table 9-6. The numbers in the columns are directional cosines squared and are referred to as loadings. We see that the first two components account for 86.13% of the explained variance of Q_m. It was further observed that the four above mentioned variates in component I identify as a size factor and that component II identifies as a slope factor.

TABLE 9-5 Eigenvalues and eigenvectors of Example 9-3

| Variables[a] | Eigenvectors | | | |
	(1)	(2)	(3)	(4)
x_1	0.895	0.356	0.093	-0.023
x_2	0.942	0.277	0.039	-0.034
x_3	-0.732	0.443	0.237	-0.164
x_4	-0.537	0.784	-0.059	0.048
x_5	0.303	-0.631	0.063	0.622
x_6	-0.200	0.939	0.069	0.010
x_7	-0.219	0.877	0.038	0.174
x_8	-0.542	0.343	0.542	0.400
x_9	0.940	0.243	0.094	0.037
x_{10}	0.939	0.263	0.072	0.057
x_{11}	-0.036	-0.652	0.619	-0.323
x_{12}	0.600	0.708	0.185	-0.124
Eigenvalues	5.12	4.24	0.81	0.73

[a] The variables are the same as in Table 9-4.

TABLE 9-6 Percent variance of 12 variates explained by each eigenvector (component) in Example 9-3

Variables[a]	Component				
	I	II	III	IV	Total
x_1	80.10	12.67	0.86	0.05	93.68
x_2	88.74	7.67	0.79	0.12	97.32
x_3	53.87	19.62	5.62	2.69	81.80
x_4	28.87	61.46	0.35	0.23	90.91
x_5	9.10	39.82	0.39	38.69	88.00
x_6	4.00	88.17	0.47	0.01	92.65
x_7	4.79	76.91	0.14	3.03	84.87
x_8	29.38	11.76	29.38	16.00	86.52
x_9	88.36	5.90	0.88	0.14	95.28
x_{10}	88.17	6.92	0.52	0.32	95.93
x_{11}	0.13	42.51	38.32	10.43	91.39
x_{12}	36.00	50.13	3.42	1.54	91.09

[a] The variables are the same as in Table 9-4.

The following decisions were made:

(1) Use only variate x_{10} in ζ_1 since it explains 88% of the information contained in that component and is by far the easiest of the four interrelated variates to measure.

(2) Use only variates x_6 in ζ_2 since it explains 88% of the information contained in that component.

(3) Delete all remaining components because the marginal variance that they explain was deemed insignificant.

Upon making these decisions, the reformulated model was

$$Q_m = ax_6^g \, x_{10}^k \qquad (9\text{-}55)$$

and components regression resulted in an overall model correlation of 0.895 as compared to 0.97 with all 11 of the variates. Hence, a reliable predictor of mean annual flood in New England was derived in terms of two very easily obtainable variates. This example was intended to illustrate how components analysis operates as a filter of redundant information and as a mechanism for model building.

Before leaving least squares and components analysis, it should be recognized that regression, as a parameter optimization technique, is limited to rather simple algebraic models, e.g., recursive relations, must be optimized by more powerful procedures. Hence, again it is emphasized that model structure and parameter optimization must be compatible.

9-5   Stepwise Multiple Regression

An alternative technique to components regression for filtering out redundant variates is stepwise multiple regression (SMR). The procedure begins with a correlation of the independent variate with the highest simple correlation with the dependent variate. The variate with the next highest simple correlation is then added and the model correlation is repeated. This process continues until the modeler decides that the marginal increase in model correlation is insignificant and the remainder of the variates are deleted. An example of this procedure has been well documented by Thomas and Benson [6] in an attempt to correlate streamflow with drainage-basin characteristics.

This procedure can filter out variates, but the objective in using SMR is not the same as that of using components regression. In SMR, there is still a strong desire to obtain correlation and there is little willingness to sacrifice it for an improved model structure; whereas in components regression, much emphasis is placed upon model structure and there is a willingness to sacrifice more correlation to achieve this goal than in use of SMR.

9-6   Rosenbrock's Method

A method of parameter optimization developed by Rosenbrock [7] has been utilized extensively in the US Geological Survey model [8]. The method is useful where the objective function of the model has partial derivatives of Q with respect to x_i, or c_i cannot be stated analytically in forms which lend the model amenable to optimization by regression techniques.

The optimization consists of a search in an m-dimensional space for the best set of m model parameters formed by m-orthogonal parameters and bounded by limits set on the parameters. The method is recursive in that it makes this search in a series of repetitive stages. Each stage is ended by evaluating a new set of m-orthogonal directions for which to search during the next stage. The new directions are based upon movements along the m-directions of the current stage. As in components analysis, only in the first search stage do the orthogonal directions coincide with the m-parameter axes. The first component of the new directions lies along the direction of fastest advance in latter stages.

Movement in each stage is made along each orthogonal direction in a series of steps. Initially, a step of arbitrary length is attempted and this move is considered successful if the new value of the objective function is an improvement over the previous value. If successful, the step size is multiplied by a factor greater than one; if unsuccessful, the move is not accepted and the step size is multiplied by a factor less than one. A new move is then attempted. The parameters are deemed optimized when at least one successful move followed by one unsuccessful move in each of the m-directions has been achieved. But in order for the optimization to be acceptable, there must be a successful move in each of the m-directions since the step size will become so small after repeated failures that it will cause no change in the objective function.

Examples in the application of the US Geological Survey model are shown in Chapter 12.

9-7   Pattern Search

A search technique [9] similar to that of Rosenbrock [7] has been computerized and has been widely used by the Tennessee Valley Authority [10, 11]. It was reported that their pattern search technique consistently reached an optimum for a variety of runoff models and data sets. This is done by finding and following a topologic feature of a map of the objective function for the range of values of the parameters. This is referred to as the "response surface" and is a hyperspace in multiple dimensions.

A unimodal response surface has only one minimum or maximum whereas a multimodal surface may have minima or maxima. No trouble has been encountered in using pattern search in optimization of multimodal surfaces, however the space can be restricted by limiting the parameter values.

Nonlinearity can severely distort the response surface to such an extent that the nonlinear least squares cannot converge to an optimum. Such distortion is termed nonconvexity. A convex space exists when a straight line connecting any two points in the space will lie entirely within the policy space.

The "pattern" in pattern search is generated by the successive points located on the response surface by sequential sets of parameter values. An exploration is conducted about each point to locate the next point. A step size is selected for each parameter and a vector is designated to perturb each parameter individually and sequentially in both a positive and negative direction. If the new point improves the objective function over the previous point a move is made in a new exploratory direction.

The effectiveness of the pattern search technique in convergence is influenced by

(1)   the initial estimates of the parameters,
(2)   the selected individual step sizes,
(3)   the speed at which step sizes are reduced, and
(4)   the convergence level specified.

Hence, there are several inputs to the computer routine which must be experimented with for the purpose of developing efficiency in parameter optimization.

Pattern search moves in the direction of steepest descent in a minimization problem and steepest ascent in a maximization problem. It can be shown that this direction is defined by the gradient vector.

The directional derivative of the vector $\mathbf{x}$ at the point in hyperspace $\mathbf{x}_0$ is

$$Df(\mathbf{x}_0) = \overline{\nabla} f(\mathbf{x}_0) \tag{9-56}$$

where l is a directional cosine and

$$|\mathbf{l}| = 1 \tag{9-57}$$

In component form, Eq. (9-56) can be written as

$$Df(\mathbf{x}_0) = \sum_{i=1}^{p} \frac{\partial f(\mathbf{x}_0)}{\partial x_i} l_i \tag{9-58}$$

To find the direction $\mathbf{l}$ such that the rate of change of $f(\mathbf{x})$ at the point $\mathbf{x}_0$ is a maximum, we maximize

$$\sum_{i=1}^{p} \frac{\partial f(\mathbf{x}_0)}{\partial x_i} l_i$$

subject to the constraint of required orthogonality

$$g(\mathbf{l}) = \sum_{i=1}^{p} l_i^2 = 1 \tag{9-59}$$

and as in components analysis we form the Lagrangian

$$F = \sum_{i=1}^{p} \frac{\partial f(\mathbf{x}_0)}{\partial x_i} l_i + \lambda \left(1 - \sum_{i=1}^{p} l_i^2 \right) \tag{9-60}$$

Differentiating Eq. (9-60), we obtain

$$\frac{\partial F}{\partial l_i} = \frac{\partial f(\mathbf{x}_0)}{\partial x_i} - 2\lambda l_i = 0 \tag{9-61}$$

and

$$\frac{\partial F}{\partial \lambda} = 1 - \sum_{i=1}^{p} l_i^2 \tag{9-62}$$

Combining Eqs. (9-61) and (9-62), we obtain

$$1 - \sum_{i=1}^{p} \frac{1}{4\lambda^2} \left(\frac{\partial f(\mathbf{x}_0)}{\partial x_i} \right)^2 = 0 \tag{9-63}$$

Solving for the eigenvalue from Eq. (9-63),

$$\lambda = \pm \tfrac{1}{2}|\overline{\nabla} f(\mathbf{x}_0)| \qquad (9\text{-}64)$$

and substituting Eq. (9-61) into (9-64) produces

$$\mathbf{l} = \frac{\pm \overline{\nabla} f(\mathbf{x}_0)}{|\overline{\nabla} f(\mathbf{x}_0)|} \qquad (9\text{-}65)$$

The plus sign produces the rate of maximum increase of $f(\mathbf{x})$ at $\mathbf{x}_0$ and the minus sign produces the rate of minimum decrease at the same point. Hence, it has been shown that the gradient vector is the direction of maximum increase of the function $f(\mathbf{x})$ because as shown in Fig. 9-5, this direction is merely a vector sum of $\partial f / \partial x_i$.

Green [9] found his pattern search version superior to the other methods previously described because of its high computational efficiency as measured by its rapid convergence and for its ability to perform optimization in multimodal space. Part of the computer package is an initial generation of the policy space as set by the limits of the parameters.

Example applications of the TVA models utilizing this pattern search routine are shown in Chapters 10 and 11.

9-8 Summary

The parameter optimization scheme chosen for a model depends heavily upon the model structure itself. This means that the model and the parameter optimization scheme chosen must be computationally compatible.

Linear least squares regression models are very easy to optimize but are so simplified that they seldom can give a high level representation of the runoff process. Further, hydrologic and physiographic variates are usually interrelated and without a statistical filtering device, the regression results in a "black box" model. These problems can be offset to some extent by utilizing the combination of nonlinear least squares, which permits a more

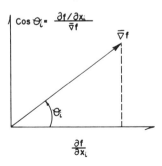

FIG. 9-5. Direction of maximum increase of objective function in pattern search.

sophisticated conceptual model of the process and components analysis, which can efficiently filter out statistically redundant variates. Although some correlation is sacrificed with this move, nevertheless, model structure is improved and regionalization of model parameters may be permitted. Stepwise multiple regression does filter out some redundant variates but many can remain. This procedure is perhaps a compromise between multiple regression and components regression.

Search techniques such as that of Rosenbrock [7] and pattern search [9] are much more powerful than nonlinear least squares. Optimization in multimodal space can be efficiently achieved and model structures which do not have analytical partial derivatives of discharge as a function of the parameter can be optimized. The latter is a strict requirement of nonlinear least squares. However, if significant interrelationships exist among the independent variates or parameters, there is little hope of filtering them out with the search routines.

Problems

9-1. Distinguish between a parameter and a variable.

9-2. What is a system constraint?

9-3. Why is there a need for an objective best fit criteria in parametric modeling?

9-4. Name at least three factors which affect the values of the optimized parameters of a parametric model.

9-5. On what does measurement of data operated upon by parametric models depend?

9-6. What potential built-in correlations can be induced in model optimization?

9-7. What is (a) unimodal, (b) multimodal policy space?

9-8. What is hyperspace?

9-9. What factors influence the structure of a parametric stormwater model?

9-10. Is global optimization of a stormwater model unconditionally guaranteed? Explain.

9-11. What effect do principal components analysis coupled with regression normally have on reducing the overall model correlation.

9-12. Is principal components analysis only a supplement to regression and not a substitute for it? Explain.

9-13. Is principal components analysis an alternative to the method of differential correction?

9-14. Are the eigenvectors produced by components analysis interrelated?

9-15. In principal components analysis, what is the sum of all of the eigenvalues equal to?

9-16. What is the statistical significance of the eigenvalue, i.e., what is it?

9-17. What is the statistical significance of the directional cosines?

9-18. What does "principal" in principal components analysis signify?

9-19. Principal components analysis is said to be an *orthonormal* system. Why?

9-20. Are the original variates related to the components? If so, how?

9-21. What is the statistical significance of the directional cosines squared?

9-22. For a given data set, how can you check that the components are not related?

9-23. Briefly and to the point, describe what principal components analysis does.

9-24. Distinguish between a linear and nonlinear system. Base your answer upon something besides mathematics.

References

1. Neilson, K. L., "Methods in Numerical Analysis." Macmillan, New York, 1957.

2. DeCoursey, D. G., and Snyder, W. M., Computer-oriented method of optimizing hydrologic model parameters, *J. Hydrol.* **9**, 34–56 (1969).

3. Kendall, M. G., "A Course in Multivariate Analysis." Hafner, New York, 1957.

4. Snyder, W. M., Some possibilities for multivariate analysis in hydrologic studies, *J. Geophys. Res.* **67**, No. 2, 721–729 (Feb. 1962).

5. Wong, S. T., A multivariate statistical model for predicting mean annual flood in New England, *Annals Amer. Assoc. Geographers,* **53**, No. 3, 298–311 (Sept. 1963).

6. Thomas, D. M., and Benson, M. A., Generalization of streamflow characteristics from drainage-basin characteristics, US Geological Survey, Open-file Report, Washington, D.C., 1969.

7. Rosenbrock, H. H., An automatic method of finding the greatest or least value of a function, *The Computer J.* **3**, 1960.

8. Dawdy, D. R., and O'Donnell, T., Mathematical models of catchment behavior, *J. Hydr. Div., Amer. Soc. Civil Engr.* **91**, HY4, (July 1965).

9. Green, R. F., Optimization by the pattern search method, TVA, Div. of Water Control Planning, Knoxville, Tennessee Res. Paper No. 7, Jan. 1970.

10. Betson, R. P., Upper Bear Creek experimental project—a continuous daily streamflow model, TVA, Div. of Water Control Planning, Knoxville, Tennessee, Feb. 1972.

11. Ardis, C. V., Jr., Storm hydrographs using a double-triangle model, TVA, Div. of Water Control Planning, Knoxville, Tennessee, Jan. 1973.

Chapter 10 | Evaluation of Effects of Urbanization and Logging on Stormwater

10-1 Introduction

Although thousands of stream gauges have been in operation in this country for decades, there is often little streamflow data available for small watershed planning applications. The reason is that water resource planning has focused largely upon basins hundreds of square miles in size. Today a considerable amount of planning is involved in smaller drainage basins such as in urban development and the 208 sections of PL 92-500, and in flood plain zoning. In the smaller drainage areas the effect of land use, soils, and physiographic characteristics upon stormwater is profound. Further, there are a multitude of these smaller basins to be considered, and the probability of finding a gauge on the stream involved or one like it is very small.

The paucity of data and lack of modeling effort has stagnated our limited knowledge of the hydrology of smaller basins and has left us with the inability to transfer stormwater information from one basin to another. Such a transfer cannot be done until stormwater response at gauged basins is related to the characteristics of each watershed. Once this is done, each stream gauge becomes a part of the statistical sample of hydrologic responses for the watersheds in the sample. Each additional watershed added to the sample provides more information about the relationships involved. The regionalization effort is seen to be a continuing effort and when it is substantially complete, it becomes possible to draw an inference of stormwater response on areas where data are not available and simulations can be reliably made.

Ardis [1] developed a model to compute storm hydrographs at gauged or ungauged sites in the Tennessee Valley from rainfall data and watershed characteristics. The model uses a unit response function to represent the response of a watershed to a given storm. The unit response function is a quadrilateral that can be formed by adding together two triangles. It is referred to herein as the TVA stormwater model. The shape of the response function is very flexible and allows the model to meet the response shape-characteristics of most of the storms analyzed.

Four parameters are needed to define the TVA stormwater model. The parameters have significant variation both within and among the watersheds studied. This response variation is found to be nonlinear and significantly related to storm and watershed characteristics. The model parameters have been regionalized and these relations are presented in Chapter 13. Ardis' data base consisted of 11 small watersheds and the model was successfully tested on independent data. Since it allows for response variation within and among watersheds, it can be used as a planning tool. To this end, the TVA stormwater model can be used on a regional basis for evaluating the effects of land use changes on stormwater response on small watersheds. Example simulations are shown in this chapter of the effects of forest cutting and urbanization on stormwater.

10-2 TVA Stormwater Model

Ardis [1] recognized that the choice of a model depends upon the objectives it intends to fulfill while living with the limitations of data reliability. The model should be only as complicated as needed to solve the problem and satisfy the objectives. An overly complex model needlessly complicates the problem thus adding to the analysis load and required input. This would inhibit its prospective use, while too little detail in the model may not yield satisfactory results. This is another way of explaining the "trade-off diagram" shown in Chapter 1. Furthermore, a large number of model parameters will result in a higher risk that a unique solution will not be obtained in parameter optimization.

In the data set available for calibrating the TVA stormwater model, only infrequently could one rain gauge per 20 sq miles be found. Hence, rainfall was assumed uniform over the basins studied and the system was considered to be lumped rather than distributed.

It was also found that the watershed systems studied were characterized by long term rather than short term time variance. This means that the response function could be considered to be time invariant throughout a storm but may be time variant from storm to storm.

It is well known from surface water hydraulics and experimental hydrology that stormwater response varies during a storm and is nonlinear. Ardis [1] attributes this effect mostly to the partial area runoff concept which was discussed in Chapter 1 rather than due to the dependence of the response function on input intensity. However, in regionalizing the stormwater model parameters, Ardis [1] did recognize a significant nonlinear system effect due to rain excess intensity. The TVA stormwater model was conceptualized to account for a quick and a delayed response which characterized the partial area contribution effect.

Wanting to keep the shape of a triangle for stormwater response and to incorporate a quick and delayed response, Ardis [1] developed a double-triangle unit response function. This was based upon the concept that the heaviest runoff into the stream is derived first from the riparian wet areas and that other areas contribute later as their soils become saturated. At the same time, the riparian areas grow in size. This concept results in an initial response and a delayed response that together form a unit response function (urf) for a given storm and basin system.

This double-triangle response is represented in Fig. 10-1. It was assumed that the delayed response peaks where the initial response ends and that both responses begin at the same time. The resulting four parameter urf is found by superposition and is shown in Fig. 10-2.

The symbols used in Figs. 10-1 and 10-2 are

I Precipitation excess intensity (in./hr). Since the volume of input is 1 basin-in., $I = 1/DT$, where

DT Time interval (hr) used for abstracting rainfall and discharge records.

C Dimensionless multiplier of I related to UP, the ordinate of the double-triangle model at $T1$. It was chosen as such for its similarity to the C in $Q = CIA$ of the "rational method."

UP Ordinate of double-triangle model at $T1$, generally the peak (in./hr).

$T1$ Time to peak of initial response (hr).

$T2$ Time base of initial response = time to peak of delayed response (hr).

$T3$ Time base of delayed response = time base of double-triangle model (hr).

R Dimensionless multiplier of DT to equal the time of peak of the initial response, $T1$.

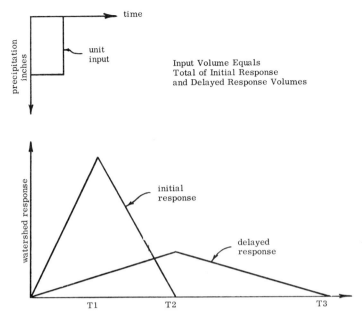

FIG. 10-1. Partial area runoff concept represented by an initial and delayed response. (After Ardis [1].)

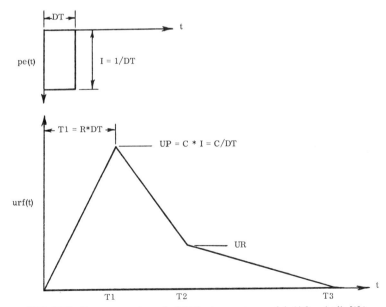

FIG. 10-2. Four parameter urf of TVA stormwater model. (After Ardis [1].)

pe(t)   Precipitation excess as a function of time t (in./hr).
urf(t) unit-response function ordinate as a function of time t (in./hr).

The basic four parameters defining the double-triangle model for a unit response function are UP, $T1$, $T2$, and $T3$. Each parameter measures a specific attribute of the model. To maintain unit volume, the variable UR is determined from the four basic parameters:

$$UR = \frac{2 - (UP * T2)}{T3 - T1} \qquad (10\text{-}1)$$

Since DT was selected to equal one hour in Ardis' [1] study, C is equal to UP and R is equal to $T1$. Also, all unit response functions described hereafter are equivalent to one-hour unit hydrographs.

Figure 10-3 is an example of the double triangle's flexibility for a constant time base $T3$. It can assume most conceivable shapes and can be "fitted" to them to approximate response behavior.

Once rainfall excess has been determined, as will be described later, a urf is derived from the observed storm hydrograph by the matrix inversion

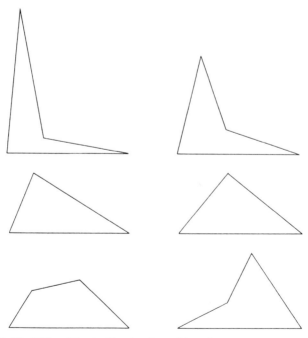

FIG. 10-3. Flexibility of the double-triangle model used as a unit response function. Time base $T3$ is constant. (After Ardis [1].)

technique of Snyder [2]. This technique has also been successfully used by Newton and Vinyard [3], and Overton [4].

Base-flow separation was employed to differentiate between "fast" and "slow" response and to eliminate the slow response. These are not the same as the initial and delayed response described earlier. Fast response corresponds with the rapid stormflow associated directly with the storm rainfall as opposed to the attenuated recessional flow from saturated soils that typically occurs several days following the storm rainfall.

Ardis [1] evaluated 12 different base-flow separation technique shapes. He found that methods other than a single straight line between point of rise and end of fast response or two straight-line segments to remove an antecedent recession showed little advantage. Although Hewlett and Hibbert [5] found that variations in separation criteria had little effect on response characteristics, their work indicated that any technique used must be reasonable and consistent since it was found that selection of the point of rise and end of fast response has a very sensitive effect on the resulting stormwater hydrograph.

Fast response ends where contributions to the total hydrograph normally considered as direct surface runoff are no longer represented at the watershed outlet. On the TVA study watersheds, this point was selected where the rate of change in total discharge became essentially constant. Selection of this point varied as much as 12 hr for the size of watersheds in Ardis' study. However, corresponding volume estimates remained below a maximum difference of 10%. A typical base-flow separation technique is shown in Fig. 10-4. Except for two of the study watersheds, an average value of $T3$ (time base of double-triangle model) was then selected for each

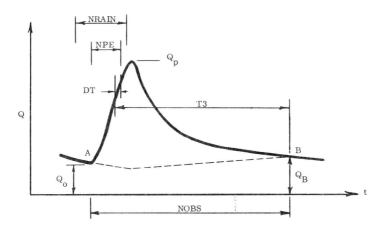

FIG. 10-4. Typical base-flow separation procedure. (After Ardis [1].)

watershed since individual differences were found to be small. Nonlinear behavior was observed during "smaller floods," hence, a constant $T3$ would not be valid. An average value of $T3$ having been selected, B was redetermined so that $NOBS$ could be determined consistent with

$$NOBS = NPE - DT + T3/DT \qquad (10\text{-}2)$$

where NPE is the number of periods, in multiples of DT, of precipitation excess estimated from the rainfall hyetograph, $NOBS$ the number of storm-hydrograph ordinates, in multiples of DT, after base-flow separation, and DT the one hour for the sample storms; therefore, $NOBS$ and NPE are in hours. In practice, the final value of $T3$ was not constant since NPE was estimated from the number of rainfall periods associated with the storm hydrograph, $NRAIN$.

With the base flow separated from the total hydrograph, the remaining volume is equivalent to the rainfall excess SRO. The volume SRO was calculated in basin inches as the area under the storm hydrograph.

The technique selected for distributing excess precipitation in time over the duration of precipitation $NRAIN$ was the Soil Conservation Service (SCS) [6] method described in Chapter 2. Although three techniques were tested, this technique was found to give consistently good results and is well known.

Mathematically, the technique reduces to

$$SRO = \frac{(ARF - IA)^2}{ARF - IA + S} \qquad (10\text{-}3)$$

where ARF is the accumulated rainfall, IA an initial abstraction from ARF, S the maximum potential retention which is related to the SCS curve number CN by definition, as $CN = 1000/(10 + S)$. The SCS technique was modified after further study showed that abstractions from hourly rainfall amounts RF_i were still too large at the beginning of storms and especially during complex storms with significant lulls which resulted in multiple-peak floods. Therefore, a constant loss parameter, PHI, for each storm was introduced to apply to each RF_i prior to accumulation over $NRAIN$ to obtain a new ARF for use in Eq. (10-3). PHI was allowed to vary from storm to storm. The new RF_i, NRF_i, is then defined as

$$NRF_i = RF_i - PHI \qquad (10\text{-}4)$$

subject to $PHI \leq RF_i$ and then

$$ARF_i = \sum^{i} NRF_i \qquad (10\text{-}5)$$

Once the S or CN for a particular storm was found, it was held constant and a vector of SRO_i values was determined from the ARF_i vector as suggested by the SCS [6]. The time-incremental values of precipitation excess PE_i were determined as in Eq. (10-6)

$$PE_i = SRO_i - SRO_{i-1} \qquad (10\text{-}6)$$

Analysis of each storm hydrograph involves determining five parameters: C, R, $T2$, $T3$, and PHI. With $DT = 1$, these are equivalent to UP, $T1$, $T2$, $T3$, and PHI, respectively, as previously described and shown in Fig. 10-2. I3 was determined by a modification of Eq. (10-2) and is an integer.

$$T3 = (NOBS - NRAIN + 1) * DT \qquad (10\text{-}7)$$

where the number of observations $NOBS$ is redefined to begin coincident with $NRAIN$. To assure that in convolution all of the double triangle is used, $T1$ and $T2$ are also required to be integers. Figure 10-5 shows how portions of the double triangle may not be used when nonintegers are used. For very peaked double triangles, the shape can be drastically modified. Since the double triangle can reduce to a single triangle to meet such a need, the following restriction was also imposed:

$$DT = 1 \leq T1 < T2 \leq T3 \qquad (10\text{-}8)$$

Storm hydrographs with distinct peaks, caused by lulls in long-duration rainfall storms, were separated during analysis and treated as separate bursts. Such complex hydrographs were separated based upon an exponential decay from an existing portion of a falling or recession limb just prior to an increase caused by the next rainfall burst. Each burst of rainfall was used with its associated portion of the complex hydrograph to evaluate the time distribution of precipitation excess for that burst. For each burst, S in Eq. (10-3) was held constant. Although S could vary among bursts,

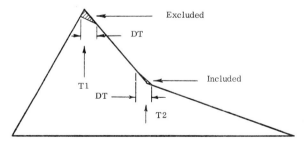

FIG. 10-5. Potential shape modification if $T1$ and $T2$ are not restricted to integers. See Eq. (10.8). (After Ardis [1].)

model parameters and *PHI* were required to be constant for all bursts in a given storm.

Initial estimates of the double-triangle parameters define a urf which, along with *PHI*, made up a set of parameters that were improved by optimization. Since linearity was assumed for each storm, the convolution of calculated precipitation excess with the double-triangle, unit response function model resulted in a predicted storm hydrograph which was compared with the observed storm hydrograph during optimization. Optimization was performed using PATSEAR, the pattern search routine of Green described in Chapter 9, and the best fit criterion used was the minimization of the weighted sums of squares of errors SSE. All errors where the observed storm hydrograph ordinates were greater than 0.1 times the maximum observed discharge were assigned a weight of 1.0; all others were assigned a weight of 0.5.

10-3   Model Tests

The model was tested on 11 small watersheds in the Tennessee Valley which had more than five years of available hydrologic data. With these restrictions and the desirability for (1) geographic and physiographic coverage across the Tennessee Valley, (2) adequate recording rain-gauge coverage, (3) stream-gauge records rated good to excellent, and with (4) minimal or no man-made regulation that affected storm hydrographs, the availability of stream gauges dropped drastically. The size range varied from 7.04 to 16.8 sq miles.

The 11 watersheds lie within the Tennessee River Basin and are shown in Fig. 10-6. Beech River and Cane Creek are in the Mississippi Alluvail Plain physiographic province of western Tennessee. Mill, Bear, Little Bear, and Whitehead Creeks, located in the Upper Bear Creek watershed in northwestern Alabama, are in the Highland Rim physiographic province. Chestuee Creek and Little Chestuee Creek are in the Valley and Ridge province in southeastern Tennessee, while the rest are in or adjacent to western North Carolina and lie in the Blue Ridge physiographic province. Basic information for each watershed is shown in Table 10-1.

Mean annual precipitation varies from 48 in. in the western portions of the sample areas to 72 in. in two of the eastern watersheds. Table 10-1 lists these data for the sample watersheds. Snowfall is generally light, and mild winter temperatures cause snow cover to persist rarely more than a few days. The mean annual snowfall varies from 3 in. in the southwestern portions to 30 in. in a small area of the mountainous northeast. The general average for the area varies from 6 to 12 in./yr. Snowmelt is not, in general, a cause of or contributor to flooding in the study region.

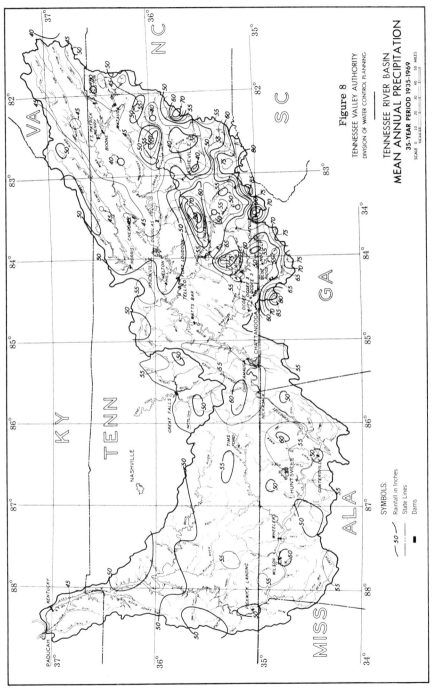

FIG. 10-6. Location map study and test watersheds. (After Ardis [1].)

TABLE 10-1 Sample watersheds[a]

Study watersheds	Assigned USGS gauge no.	Owner	Length of record	Drainage area	Mean elev.	Mean annual Precip.	Temp.	Comments
1 Crab Creek near Penrose, North Carolina	3–4420	USGS	1942–1955	10.9	2625	72	54	Mountainous with bottom agriculture
2 Boylston Creek near Horseshoe, North Carolina	3–4440	USGS	1942–1955	14.8	2425	57	55	Mountainous with considerable bottom agriculture, unusual drainage pattern
3 S. F. Mills R. at The Pink Beds, North Carolina	3–4445	USGS	1926–1949	9.99	3730	69	50	Forested mountain area
4 Allen Creek near Hazelwood, North Carolina	3–4575	USGS	1949–	144	4440	57	48	Forested mountain area
5 N. Indian Creek near Unicoi, Tennessee	3–4650	USGS	1944–1958	15.9	3095	53	52	Mountainous with bottom agriculture
6 Noland Creek near Bryson City, North Carolina	3–5135	USGS	1935–	13.8	4050	68	49	Rugged forested mountain area
7 N. F. Citico Creek near Tellico Plains, Tennessee	3–5184	TVA	1960–1968	7.04	3325	71	51	Rugged forested mountain area
8 Little Chestuee Creek below Wilson Station, Tennessee	3–5650.8	TVA	1947–1957	8.24	1035	53	58	Rough with agriculture
9 Cane Creek near Shady Hill, Tennessee	3–5944.55	TVA	1953–1964	16.8	480	49	60	Agricultural
10 Mill Creek near Bear Creek, Alabama	3–5917.80	TVA	1962–	7.18	910	57	62	Agricultural and forested
11 Bear Creek near Carroll Crossroads, Alabama	3–5915.40	TVA	1963–	13.9	990	54	62	Forested area

[a] After Ardis [1].

The remnants of hurricanes, frequent severe thunderstorms during the warm season, and warm frontal rains during the winter produce the flooding (storm hydrographs) experienced in the Tennessee River Basin. Most rainfall stations have experienced 24-hr amounts in excess of 6 in. in the spring, midsummer, and fall. The winter season, December through March and April, is the wettest period of the year. This period is frequently subjected to large-scale weather systems characterized by surface frontal activity (generally quasi-stationary), large troughs of low pressure above the surface, and a rich supply of moist air from either the Gulf or the Atlantic. During the period 1936 to 1969, the maximum recorded rainfall in 1 hr was 5.50 in., in 3 hr, 8.80 in. in 6 hr, 11.13 in., and in 24 hr, 13.10 in.

The sample watersheds are essentially devoid of lakes and swamps; however, a few man-made ponds are present in some watersheds to a minor and insignificant degree. Karst topography is present in those watersheds in the Ridge and Valley province (Chestuee Creek and Little Chestuee Creek), but its effect is not yet known. Sinkhole areas are small and constitute less than 5% of the area; the Chestuee Creek watershed is affected the most, Little Chestuee Creek a negligible amount. Sinkhole areas were eliminated from consideration when encountered by deleting associated drainage areas from watershed characteristic analyses.

Table 10-2 compares the correlation coefficients obtained during analysis for each of the 140 storms. On the average, 96% of the variance, ranging from low to high of 90 to 99%, was explained. There were a few outliers, the worst explaining only 82%. Figures 10-7–10-9 compare fair, average, and good model fittings.

It is quite evident from Table 10-2 that a wide range of storms was studied. For example, stormwater peak discharge (QMAX) ranged from 173 to 4929 cfs over all 140 storms. On a given watershed, the ratio of maximum to minimum QMAX ranged from 1.4 to 26. Storm durations lasted from one hour to more than one day, and total precipitation ranged from less than 0.5 in. to nearly 9 in. Precipitation excess volumes covered a range from 0.14 to 3.43 in. Typical extremes of model parameters varied by an order of magnitude over the 140 storms.

Table 10-3 summarizes the within- and among-watershed variation of the 140 storms. Mean storm attributes and stormwater model parameters are listed with the coefficient of variation CV shown for each by watershed. $T3$ is the only variable that does not exhibit large within-watershed variation.

The variation of the parameters was expressed algebraically in terms of selected watershed and rainfall characteristics. These results, which are presented in Chapter 12 comprise the present regionalization of the model and provide the basis for parameter prediction and stormwater simulation on ungauged watersheds.

TABLE 10-2   Correlation coefficients from analysis[a]

Watershed	No. of storms	Correlation coefficient[b]		
		Lowest	Highest	Average
Crab	6	0.967	0.997	0.985
Boylston	8	0.974	0.995	0.984
South Fork Mills	6	0.953	0.990	0.976
Allen	4	0.977	0.991	0.982
North Indian	5	0.904	0.992	0.966
Noland	6	0.918	0.996	0.969
North Fork Citico	12	0.948	0.988	0.972
Little Chestuee	9	0.980	0.998	0.991
Cane	26	0.952	0.993	0.980
Mill	26	0.930	0.996	0.979
Bear	32	0.934	0.997	0.979
	140	Storm Average		0.980
Highest		0.980	0.998	0.991
Lowest		0.904	0.988	0.966
Average		0.949	0.994	0.978

[a] After Ardis [1].

[b] Simple correlation coefficient after optimization based on weighted fit during optimization between observed and predicted hydrographs.

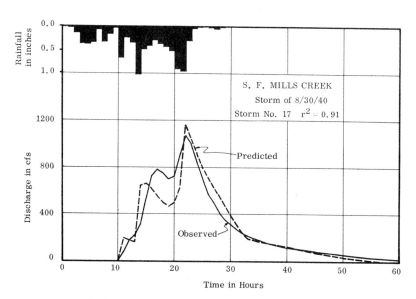

FIG. 10-7. Example of fair fit model. (After Ardis [1].)

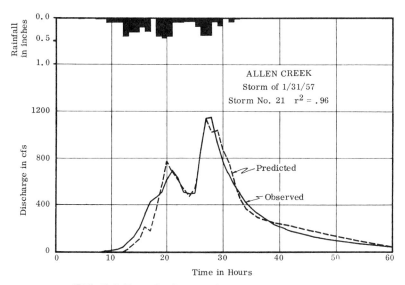

FIG. 10-8. Example of average fit model. (After Ardis [1].)

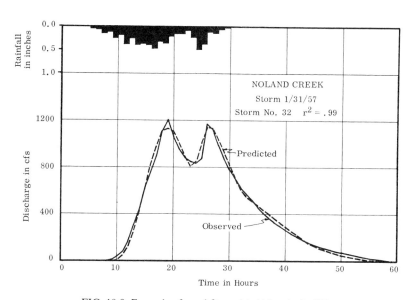

FIG. 10-9. Example of good fit model. (After Ardis [1].)

10-3 Mean data for storm attributes and double-triangle parameters from analysis of study watershed hydrographs[a]

| Study watershed | No. of storms | Qmax/Area | | Storm attributes | | | | | | Double-triangle parameters | | | | | |
| | | | | SRO | | NPE | | UP | | T1 | | T2 | | T3 | |
		Mean	CV[b]	Mean	CV	Mean	CV	Mean	CV	Mean	CV	Mean	CV	Mean	CV
Crab	6	74.7	47	0.884	37	7.3	30	0.201	19	3.00	28	6.5	24	26.0	6
Boylston	8	32.4	39	0.841	38	10.6	34	0.059	29	2.37	75	14.4	37	26.1	13
South Fork Mills	6	91.4	64	1.746	55	9.7	41	0.147	41	2.83	58	9.0	33	39.0	5
Allen	4	84.6	15	1.221	31	9.7	43	0.207	20	1.75	26	5.5	28	43.3	3
North Indian	5	25.1	12	0.514	58	12.4	75	0.150	48	3.80	56	8.6	26	33.8	5
Noland	6	86.6	9	1.252	58	9.2	63	0.173	34	1.83	39	6.8	33	30.3	5
North Fork Citico	12	101.8	28	1.909	27	9.5	41	0.107	21	4.00	40	12.4	26	43.6	4
Little Chestuee	9	84.3	33	1.127	38	8.1	37	0.187	16	2.78	34	8.2	20	22.9	8
Cane	26	75.7	38	1.053	68	6.6	63	0.156	25	3.96	33	11.5	27	40.2	6
Mill	26	135.8	94	1.170	63	5.4	74	0.215	21	2.81	30	7.1	18	22.1	23
Bear	32	62.1	103	1.188	69	8.7	68	0.084	39	7.75	26	16.8	24	58.0	26

[a] After Ardis [1].
[b] CV = coefficient of variation.

10-4 Simulated Effects of Forest Cutting on Stormwater

EXAMPLE 10.1 North Fork Citico Creek is a very steep mountainous watershed completely in forest. It will be assumed that the area will have its timber harvested and that those areas harvested are replanted with seedlings. During the first few years before the watershed recovered from the cut, its response for various percentages of the area in forest would be as shown in Fig. 10-10. For this winter storm it was assumed that the watershed would be in such a condition that for each percentage of the area in forest the runoff volume remained the same. The observed and simulated storm hydrograph for 100% forest is shown for a comparison check. The time to peak discharge would decrease and the magnitude of peak discharge would increase as more trees are removed. However, the change is not linear. The rate of change of peak discharge increases more rapidly when the percent area in forest is less than 50% than when it is between 50 and 100%. Therefore, when a 10% forest cut is planned, it would be noted that the hydrologic effect would be greater if the present condition were 25% than if it were 75% in forest. From a qualitative point of view, this general response behavior may have been anticipated for a forest harvest, but with the aid of the TVA stormwater model it can be quantified.

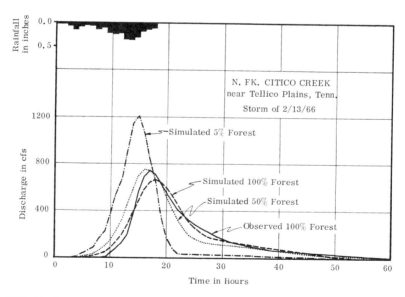

FIG. 10-10. Simulated effect of forest cutting at North Fork Citico Creek watershed for storm of February 13, 1966. (After Ardis [1].)

EXAMPLE 10.2   Bear Creek watershed is cited as another example of the potential effect on response behavior due to land-use change. It is presently 85% forest and 15% agriculture. However, the soils and topography of the watershed are such that these two uses could be interchanged. The regionalized TVA stormwater model can be used to project the effects of this proposed change.

The storm of December 22, 1968 was used as an example. Rainfall was 2.67 in. with an observed storm-runoff volume of 1.01 in. The continuous daily-streamflow model [8] was used to simulate a storm-runoff volume of 0.98 in. under existing conditions but 1.68 in. with only 15% of the area in forest. By using a *SRO* of 1.68 in. from the same simulated rainfall of 2.67 in., the stormwater model indicated that the watershed would be considerably more flashy.

The observed and simulated storm hydrographs for existing and reversed land uses are shown in Fig. 10-11. Not only did the change in land use cause the watershed to be more responsive, but it also increased the storm runoff from the same rainfall. The combined effect produced a flood peak over 2.5 times that under existing conditions. Further, most of the runoff volume would have already left the watershed at a time where under existing conditions it would just be reaching its peak. It is evident from Fig. 10-11 that

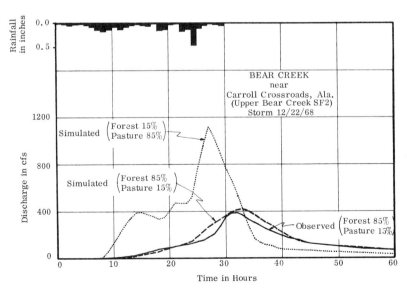

FIG. 10-11. Simulated effect of forest cutting at Bear Creek watershed for storm of December 22, 1968. (After Ardis [1].)

if the existing land uses were interchanged, we would have to plan to manage the increased volumes and rate of runoff.

10-5   Simulated Effects of Urbanization on Stormwater

EXAMPLE 10.3   The West Town tributary to Fourth Creek, Knoxville, Tennessee is a 0.82 sq. mile watershed which has undergone substantial development in the last 5 to 10 yr. (See Fig. 10-12.) The watershed is basically wornout farm land with soils in the SCS C-group. In the past decade, 1965–1975, a major shopping center has been built in the headwaters and a 72-in. storm sewer system was installed beneath the parking lot. This was modeled in Chapter 6. Restaurants, shops, and quick food establishments now line both sides of Kingston Pike. Single family residences and garden apartments account for the remaining development. The watershed is about 100% developed and is 60% impervious.

The Tennessee Valley Authority has operated a continuous streamgauge on the West Town tributary since early in 1971. It was installed about the time ground was broken for the new West Town shopping center. At this

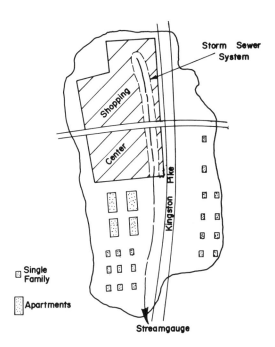

FIG. 10-12. West town tributary to Fourth Creek, Knoxville, Tennessee.

writing, some of the stormwater events have been analyzed using the TVA stormwater model. The results have shown that the basin is very flashy. As little as 0.05 in. of rain has produced peak flows at the streamgauge of 60 ft³/sec or about 73 ft³/sec/sq mile!

High quality data for eight storms was analyzed for the period of spring and summer of 1972. This was completed after the West Town Mall was completed and the watershed was essentially 100% developed. The storms were all short duration with fairly intense rainfall. A DT of 5 min was used for all storms and the eight unit response functions derived from the stormwater hydrographs are shown in Fig. 10-13. At a glance the results seem to widely vary. The optimized parameters and fixed statistics are shown in Table 10-4. The optimized SCS CN lie between 85 and 100 indicating in some instances that a 100% runoff condition occurs.

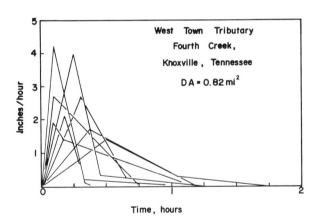

FIG. 10-13. Five-minute response functions, April through September 1972.

TABLE 10-4 Optimized parameters in TVA stormwater model applied to storms in West Town Tributary, Knoxville, Tennessee

Storm date	T1 (min)	T2 (min)	UP (in./hr)	PHI (in./5 min)	CN	Coef. of Determination
4/20/72	5	10	2.76	0.025	95	0.901
5/3/72	5	10	1.91	0.036	100	0.835
6/6/72	30	40	2.81	0.001	86	0.990
7/29/72	10	40	2.17	0.020	99	0.939
8/7/72	5	20	4.23	0.130	97	0.943
8/23/72	20	62	1.69	0.001	91	0.953
8/23/72	14	24	3.97	0.000	94	0.939

An explanation of the variation of the urf from storm to storm will be attempted. Time to peak of the urf is plotted versus the associated storm average rain excess intensity in Fig. 10-14. This follows the same trend as the variation of lagtime or time of concentration with supply rate derived from the unsteady hydraulic equations in Chapter 4.

Lagtime for each storm was calculated from each urf and divided by 0.6 to form an estimate of time of concentration which was plotted versus the generating rain excess intensity in Fig. 10-15. The data are approximated by

$$t_c = 13/i_e^{0.4} \qquad (10\text{-}9)$$

and the value 13 is an experimental determination of the lag modulus. Neglecting time of concentration of the sewer and channel and considering the West Town tributary as a V-shaped watershed, lag modulus of the basin is approximately

$$13 = 0.928\left(\frac{nL}{\sqrt{S_0}}\right)_0^{0.6} \qquad (10\text{-}10)$$

The average length of each plane is 1520 ft and the average slope is 0.036. The average n-value of each plane can be approximated from Eq. (10-10).

$$n = (13/0.928)^{1.67}(\sqrt{S_0}/L_0) = 0.010 \qquad (10\text{-}11)$$

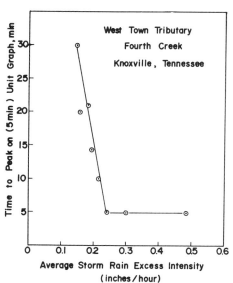

FIG. 10-14. Time to peak as a function of rain excess rate for summer storms of 1972.

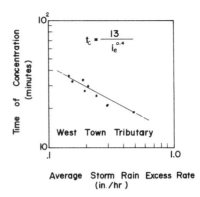

FIG. 10-15. Time of concentration as a function of rain excess rate for summer storms of 1972.

This *n*-value corresponds to smooth concrete or tin roofs and gives added proof that the watershed is very flashy.

This example has also been used to demonstrate how the parametric and deterministic models can be linked. This linkage provides an improved notion of the process, and strengthens the TVA stormwater model as a predictor by providing a physical explanation of the model parameters.

EXAMPLE 10.4   One of the storms analyzed in Example 10.3 occurred on September 1, 1972. Approximately 0.63 in. of rain fell in 15 min, and 0.039 in. of effective rainfall was generated producing the basin storm hydrograph shown in Fig. 10-16. Also shown is the fitted TVA stormwater

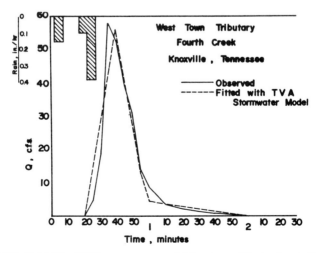

FIG. 10-16. Fit of TVA stormwater model for storm of September 1, 1972.

model. Prior to development, the *CN* for the basin was approximately 65 as estimated by the SCS procedure shown in Chapter 2. Equation (10-3) indicated that no effective rainfall would have occurred, and therefore no effective rainfall would have occurred for any of the eight storms analyzed.

10-6 Model Limitations

Presently, the TVA stormwater has applicability only in rural areas of the Tennessee Valley and any extrapolation of the model should first involve a careful examination of the watershed characteristics of the study basins. The model is not as yet ready for use as a generalized predictor of stormwater in urban, mountainous, or logged watersheds without further work.

Problems

10-1. Verify Eq. (10-1) and Eq. (10-2).

10-2. Formulate and verify your own parametric stormwater model by applying it to a watershed in your region.

10-3. Apply the TVA stormwater model to an urbanizing watershed in your region and predict the change in the stormwater regime for planned urbanization.

10-4. What reliability do you place on the simulation of forest cutting on stormwater in Example 10.1?

10-5. Apply the TVA stormwater model to a watershed in your region which may undergo a substantial cutting operation and compare your simulation results with those in Examples 10.1 and 10.2.

10-6. Criticize the attempt in Example 10.3 to link parametric and deterministic modeling of the West Town Tributary.

10-7. Apply the techniques in Example 10.3 to a watershed in your region which has undergone or has planned urbanization similar to West Town Tributary. Compare the results with those of Example 10.3.

10-8. Explain why the unit hydrograph is variable from storm to storm on a watershed which has a constant land use.

References

1. Ardis, C. V., Jr., Storm hydrographs using a double triangle model, TVA, Div. of Water Control Planning, Jan. 1973.
2. Snyder, W. M., Hydrograph analysis by the method of least squares, *Proc. Separate, Amer. Soc. Civil Engr.* No. 793 (1955).

3. Newton, D. W., and Vinyard, J. W., Computer-determined unit hydrograph from floods, *J. Hydr. Div., Amer. Soc. Civil Engr.* **93** (HY5), 219–235 (Sept. 1967).

4. Overton, D. E., A least-squares hydrograph analysis of complex storms on small agricultural watersheds, *AGU, Water Resources Res.* **4,** No. 5, 955–963 (Oct. 1968).

5. Hewlett, J. D., and Hibbert, A. R., Factors affecting the response of small watersheds to precipitation in humid areas. *In* "Forest Hydrology" (W. E. Sopper, Ed.), pp. 275–290. Pergamon, Oxford, 1967.

6. Soil Conservation Service, "National Engineering Handbook," Section 4, Washington D. C., 1964.

7. Green, R. F., Optimization by the pattern search method, Research Paper No. 7, TVA, Knoxville, Tennessee, Jan. 1970.

8. Betson, R. P., Upper Bear Creek experimental project—a continuous daily-streamflow model, TVA, Knoxville, Tennessee, Feb. 1972.

Chapter 11 | Evaluation of the Effects of Strip Mining on Streamflow

11-1 Introduction

In order to promote stripped coal as an economically and environmentally acceptable fuel source, it is essential to assess properly the external costs of extracting coal by strip mining. One of the elements contributing to the damage function associated with these costs on the production of extracting coal by means of stripping, is the impact on the watershed hydrologic environment. Two of the most important hydrologic changes in watershed response are peak and daily flow due to the potential to cause flood damages, the impact on water supply, and the general quality of receiving water bodies.

In this chapter two important studies of the effects of strip coal mining on watershed hydrologic response are presented. The TVA daily flow model has been used to evaluate the effects of surface coal mining on continuous daily flows in a 382 sq. mile watershed in East Tennessee [1] and the Stanford model has been used to evaluate the effects of surface coal mining and acid mine drainage on continuous daily flow and water quality in a watershed in Fayette County, Pennsylvania [2].

11-2   TVA Daily Flow Model

The selection of the watershed model in the study by Tung [1] was based upon the following requirements and constraints:

1. There must be a model to simulate flow on a continuous basis.
2. The model must have an objective best-fit.
3. Daily rainfall must be the input to the model.

Based on the above requirements and constraints, the TVA daily flow model [3] was selected. Other versions of models were excluded either because required data are too severe, an objective best-fit criterion is lacking, or because the model is not amenable to the analysis of continuous streamflow.

Model Development

The model is basically a simple system of water budget bookkeeping for the watershed. The incoming precipitation is distributed among a cascading series of watershed hydrologic compartments. The input to the model requires daily rainfall and streamflow and monthly evapotranspiration for analysis runs. Output from the system consists of daily, monthly, and annual streamflows. The system is schematically represented in Fig. 11-1. The model parameters and constants are listed in Table 11-1.

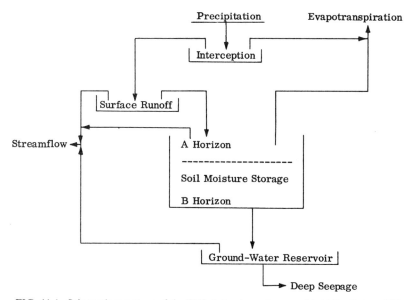

FIG. 11-1. Schematic structure of the TVA daily streamflow model. (After Betson [3].)

TABLE 11-1   TVA model parameters and constants

Primary model parameters
  1. *B*   a volumetric parameter used to preserve mass balance
  2. *A*   a surface runoff volume parameter (associated with winter storms)
  3. *D*   a surface runoff volume parameter (associated with summer storms)
  4. *GWK*   a groundwater volumetric parameter
  5. *TDSRO* a surface runoff routing parameter

Secondary model parameters that may be optimized
  1. *SROK*   a surface runoff recession constant
  2. *BHORP*   soil **B** horizon permeability
  3. *GROKW* a winter ground-water recession constant
  4. *GROKS* a summer ground-water recession constant

Model constants
  1. *ACREIN*   drainage area in square miles
  2. *WCEPT*   winter interception capacity
  3. *SCEPT*   summer interception capacity
  4. *AHORD*   moisture capacity in the soil A horizon
  5. *GWDOR*   a groundwater reservoir allocation constant
  6. FALL, WINTER, SPRING, SUMMER   day of the year for the beginning of the
                     respective season

Interception Storage.   The first of the series of compartments, interception storage, has nonparametric seasonal variations in the model. All incoming moisture enters interception storage until a preassigned volume is filled. In general, values of 0.05 and 0.25 in. for winter and summer, respectively, have been found to be reasonable for the forested watersheds in the Tennessee Valley Region.

Surface Runoff Volume.   After satisfying interception, the residual precipitation becomes potential surface runoff. The surface runoff is determined by a functional relationship between the watershed retention index and rainfall minus interception as

$$RI = [A + (D - A) * SI] \exp[-(SMI + GWR)] \quad (11\text{-}1)$$

$$SURVOL = (RF^2 + RI^2)^{1/2} - RI \quad (11\text{-}2)$$

where

  RI   retention index (in.);
  A   a parameter associated with winter storms (in.);
  D   a parameter associated with summer storms (in.);
  B   a parameter used to force continuity (in.$^{-1}$);
  SI   a season index that equals one in summer and zero in winter
        and with external variable phasing;

    SMI   the moisture (in.), stored in the soil moisture compartment;

    GWR the volume of water stored in the ground water reservoir (surface in.);

SURVOL daily surface runoff volume to be routed (surface in.);

    RF   daily rainfall minus interception (in.).

Equation (11-2) is capable of estimating lower (approaching 0%) and upper limit (approaching 100%) values of surface runoff to assure reasonable measures.

The runoff index is related to physical watershed characteristics and antecedent moisture conditions. The two coefficients A and D providing the winter and summer indexes, respectively, are parametric measures of the moisture storage capabilities of watershed soils.

The parameter B is determined in the model to conserve mass balance between the predicted and observed total runoff volumes over the period of analysis. The seasonal variable SI is associated with crop conditions and is used to differentiate between winter and summer. The value of SI is assigned to be zero in the winter, one in summer, and interpolated values in between for spring and fall.

Groundwater Runoff Volume. After interception storage and surface runoff volume have been removed, the excess of precipitation then becomes a potential for groundwater runoff. The portion of this remaining excess that becomes groundwater is assumed to be proportional to the yield of storm runoff as

$$GWV = (SURVOL * GWK/RF) * PE \qquad \text{and} \qquad GWV \leq PE \qquad (11\text{-}3)$$

where

    GWV a volume to be added to the groundwater compartment (in.^{3});

    GWK a parameter which relates the yield of groundwater for a storm to the yield of storm runoff;

    PE   the available moisture potential from precipitation after interception and surface runoff are removed (in.).

Soil Moisture Storage. The residual precipitation excess along with the groundwater volume is allocated to the soil moisture storage. This compartment is used, along with the groundwater reservoir, to define the status of moisture in the system for the surface runoff relations, Eq. (11-1). The two compartments (shown in Fig. 11-1), A and B horizons constitute the soil moisture storage. The A horizon compartment simulates the upper soil horizon and determines its capability to store and/or to percolate moisture into the lower soil zone. Depletion from this compartment occurs as evapotranspiration and as percolation to the B horizon compartment at a rate determined by the permeability of the soil. The A horizon feature is used where moisture storage in the A horizon is relatively limited and the B horizon is relatively impermeable.

Evapotranspiration. Monthly evapotranspiration values are required as input to the model. These input data are computed using the technique described in the TVA daily flow model as shown in the following:

$$(\overline{RF} - \overline{RO}) \cong \sum_{i=1}^{12} \overline{ET_i} = \sum_{i=1}^{12} K(\overline{EP} * GI)_i \qquad (11\text{-}4)$$

where

$\overline{RF}$ long-term average annual rainfall (in.);
$\overline{RO}$ long-term average annual streamflow (in.)
$\overline{ET}$ monthly evapotranspiration (in.)
K a factor which preserves mass balance of evaporation loss according to long-term records;
$\overline{EP}$ long-term average monthly pan evaporation or a measure of potential evapotranspiration (in.);
GI growth index of crop—a ratio of current evapotranspiration to that at maturity

In Eq. (11-4), a growth index of 0.5 during the winter increasing to 1.0 during the summer has been for forested watersheds.

Deep Seepage. The model was originally developed without an explicit term for deep seepage losses. For most catchment studies by TVA with the model, deep seepage was negligible. In a catchment where significant deep seepage losses do occur, these losses will be absorbed into the calculation for K in Eq. (11-4), since in this case the long-term rainfall minus runoff term would include these losses.

Runoff Routing. The daily surface runoff and groundwater runoff are determined by distributing the daily surface runoff volume from Eq. (11-2) and the daily groundwater runoff volume from Eq. (11-3) in time by routing. For surface runoff, the routing is determined as in Eq. (11-5).

$$SRO_i = TDSRO * SURVOL_i + SURES_i * (1 - SROK) \quad (11\text{-}5)$$

where

SRO the routed surface runoff (in.);
$TDSRO$ a primary model parameter;
$SURES$ the surface runoff reservoir (in.);
$SROK$ a surface recession parameter.

Equation (11-5) states that the ith day's surface runoff component of streamflow will be a percentage ($TDSRO$) of the $SURVOL$ determined from the ith day's rainfall plus the portion of that occurring from the previous rainfalls as determined by a surface recession parameter.

Groundwater is routed daily from the groundwater reservoir (GWR) by using a recession constant as in Eq. (11-6).

$$GRO_i = GWR_i * (1 - GROK) \qquad (11\text{-}6)$$

where

GRO daily groundwater runoff (in.);
GWR groundwater reservoir (in.);
$GROK$ groundwater recession constant.

Model Optimization

A modified pattern-search technique PATSEAR [4] has been used to determine the optimal set of parameter values in the model, by finding and following a topologic feature of the response surface. The response surface is the mapping of the objective function for the range of values of the parameters. The basic objective function used in the model is to minimize the sums-of-squares of the errors between predicted and observed values, i.e., daily streamflow. The advantage of the PATSEAR package has been its ability to consistently reach an optimum solution for different types of models and data sets.

A Case Study

Tennessee's coal fields are restricted to the Pennsylvanian formation on the Cumberland Plateau, which is part of the Appalachian Region, extending southwesterly across the eastern part of the State. The area covers about 5000 sq miles including all or parts of 22 counties.

The estimated recoverable strippable resources and strippable reserves of coal in Tennessee as of January 1, 1968 is vast. About 58% of the State's coal production was from strip mining, 3.4% from auger mining, and the remainder from underground mining in 1971. Approximately 70% of all strip coal mining in Tennessee has been done in the New River basin. As of 1972, 1150 linear miles had been stripped in the 382-sq mile basin.

The New River watershed, shown in Fig. 11-2, has a drainage area of 382 sq miles and is located in the southern Appalachian Region in Tennessee. It was selected for study because the watershed has been intensively (intensified locally by the form of multiseam cut) and extensively (intensified mining being carried out over the basin) mined for coal and long-term

Scale 1:250,000

FIG. 11-2. Topographic map of the New River Watershed, Tennessee.

hydrological data and other related information were available. The TVA daily flow model, which has been successfully applied to simulate daily streamflow for watersheds with various drainage areas and characteristics in the Tennessee Valley Region [3], was chosen to simulate different watershed conditions in terms of degree of mining disturbance. Four watershed study time periods, representing different accumulated mining disturbance levels and ranging from 0 to 5% of the total watershed area disturbed in the New River Watershed, were selected for analyzing progressive effects of stripping on streamflow.

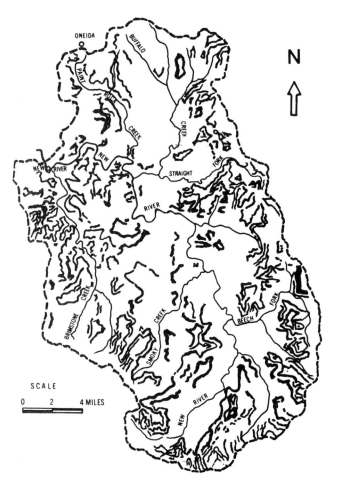

FIG. 11-3. Extent of contour strip coal mining in the New River Watershed, Tennessee from 1943 to 1973 (31, 4, 19, 44, 32).

Contour strip coal mining in the New River Watershed began in the early 1940s. By the end of 1974, an estimated 12,000 acres, or about 5% of the total watershed area, have been disturbed by contour stripping coal in the New River Watershed as shown in Fig. 11-3. From aerial photographs taken by various agencies, it was estimated that less than 1% of the total watershed area had been affected by the end of 1955; the accumulated percentage increased to about 2.5% by the end of 1965; and finally about 5% by the end of 1974.

Coal contracts for more than 5 million tons of coal from the New River area have already been awarded to 12 suppliers by TVA for the Kingston Steam Plant during the period from 1975 to 1977. About 55% of the total tonnage will come from contour strip mining operations and the rest will be obtained from underground mines. This amount of stripped coal potentially represents approximately 2500 more acres of disturbed land in this area during the period from 1975 to 1977.

At present, two rainfall stations are located within the watershed, and several others are located around the periphery. The Theissen polygon method [5] was used to average daily rainfall for five gauges for input to the model. Continuous daily discharge above the New River gauging station has been published by the United States Geological Survey in the Water Supply Papers. The average daily discharge of the watershed from October 1, 1934 through September 30, 1973 was 727 cfs.

The results of the model calibration analysis are given in Tables 11-2 and 11-3. An example calibration is plotted for 1973–1974 in Fig. 11-4. Table 11-2 gives the optimal model parameter values representing the best of the corresponding watershed conditions for the four selected watershed study periods.

TABLE 11-2   Optimal model parameter values and constants for the four selected watershed study periods

Study period (water years)	Primary model parameters*					Secondary parameter[a]	Model constants	
	A	D	GWK	$TDSRO$	B	$SROK$	$BSMI$	BGW
1970–1974	35	221	1.78	0.684	0.224	0.323	4	1
1961–1965	89	190	2.40	0.676	0.411	0.281	4	1
1951–1955	98	148	1.70	0.700	0.543	0.300	4	1
1943–1947	158	124	1.91	0.612	0.576	0.350	4	1

[a] Optimized values. All other parameter values and constants listed in the following remain the same for the four watershed periods: DA, 382; $GROKW$, 0.92; $WCEPT$, 0.05; $BHORP$, 3.0; $AHORD$, 3.0; $SCEPT$, 0.20; $GROKS$, 0.95; HOR, 0.0; $GWDOR$, 0.001.

TABLE 11-3 Correlation of the TVA daily flow model calibration for the four watershed study periods

Watershed study period	Correlation coefficient	Standard error (cfs) (in./day)	
1973–1974	0.96		
1972–1973	0.89		
1971–1972	0.92		
1970–1971	0.86		
1969–1970	0.82		
		(980)	
Overall, 1970–1974	0.90	0.094	
1964–1965	0.91		
1963–1964	0.87		
1962–1963	0.94		
1961–1962	0.95		
1960–1961	0.95		
		(576)	
Overall, 1961–1965	0.93	0.056	
1954–1955	0.94		
1953–1954	0.95		
1952–1953	0.89		
1951–1952	0.92		
1950–1951	0.93		
		(660)	
Overall, 1951–1955	0.93	0.064	
1946–1947	0.87		
1945–1946	0.89		
1944–1945	0.82		
1943–1944	0.86		
1942–1943	0.95		
		(700)	
Overall, 1943–1947	0.89	0.068	

EXAMPLE 11.1 After the completion of the model calibration, the calibrated model for the four selected watershed study periods was then used to simulate daily streamflow. A total of 20 water years of the simulated daily streamflow has been generated for comparison in terms of degrees of mining disturbance and rainfall patterns.

One example comparison is shown in Fig. 11-5. It is a comparison of the simulated water years of 1973, 1965, and 1954 representing wet, normal, and dry years, respectively. In all three cases, the simulated stormflow and

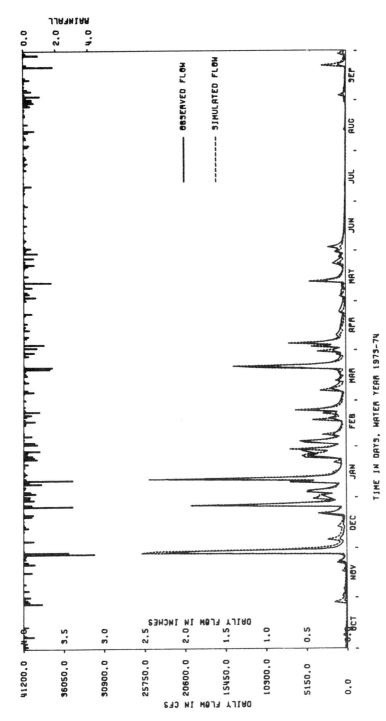

FIG. 11-4. Simulated versus observed daily streamflows, New River Watershed, Tennessee, 1973–1974.

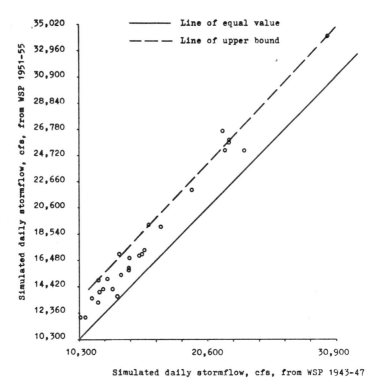

FIG. 11-5. Comparison of the simulated daily stormflows between WSP 1970–1974 and 1943–1947.

low flow increased when compared with the virgin condition. Stormflow (equaling 2% of the time or less) increased significantly, but low flow (equaling 90% of the time or more) remained relatively unchanged in the early stages of mining; when strip mining intensified and extensified over the watershed, both stormflow and low flow increased significantly, nevertheless, stormflow showed greater fluctuation than the previous stages of mining. The results were found to be statistically significant.

In general, the results indicated that contour strip mining in the New River Watershed over the last three decades has increased peak flows and low flows as well. Changes in peak flows as a result of contour mining depended on the magnitude or type of storm, time of year, and amount of stripping.

In the early stages of contour mining, peak flows increased significantly because the highwall and bench areas created by stripping were capable of intercepting and then routing directly a considerable volume of surface flow through ditches and bench areas to the stream; nevertheless, low flows remained unchanged due to the relatively small scale of bench areas and spoils for storage. As contour mining intensified, the huge dimensions of bench areas and spoils created from multiseam mining were capable of storing considerable volumes of surface water and hence peak flows decreased markedly; on the other hand, the impounded water in bench areas and spoils was gradually released to the stream as subsurface seepage which thereby increased low flows. As contour mining further extensified over the watershed, the increases in dimensions of bench areas and spoils further increased storage capacity.

Model Limitations

The TVA daily flow model has been applied to only one basin involving coal strip mining. Hence, there is a need for further testing on additional watersheds with varying characteristics. It should be recognized that the simulated changes in streamflow predicted by the model should be placed relative to the errors associated in fitting the model to the data. Otherwise, we would not know if a predicted streamflow response change was real or simply "noise" in the model or the data. Since the answer to this problem is a matter of degree, the modeler needs to be skilled to some extent in significance testing. However, this is beyond the scope of this book.

11-3 Stanford Model

Model Development

This model has had widespread application in recent years, primarily because of its versatility. Figure 11-6 is a flow chart showing the structure of the model. This model should be calibrated to each watershed from existing records of rainfall and streamflow. The only input parameters required are rainfall and potential evapotranspiration in addition to physical descriptions of the watershed and hydraulic properties.

The sequence of calculations is diagrammed in Fig. 11-6. The use of three zones of moisture regulate soil–moisture profiles and groundwater conditions. The rapid response encountered in smaller watersheds is accounted for in the upper zone, while both upper and lower zones control such factors as overland flow, infiltration, and groundwater storage. The lower zone is responsible for longer-term infiltration and groundstream flow is a combination of overland flow, groundwater flow, and interflow.

The complete description of the model is beyond the scope of this text. Applications have been made to both urban and rural watersheds and the

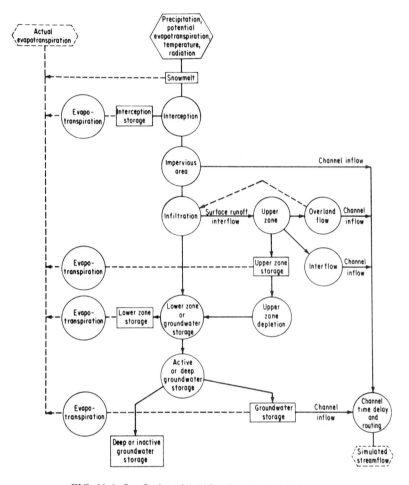

FIG. 11-6. Stanford model. (After Crawford and Linsley [6].)

results are dependent upon the care exercised in calibrating the model. In this text, a particular application to coal strip mining [2] will be presented.

The Stanford model [6] consists of a group of mathematical expressions which establish relationships and interactions between elements of the hydrologic cycle. The model is normally operated with known input and output and model parameters are optimized until an acceptable fit is achieved between estimated and actual discharge values. Input can be classified in six groups:

(a)   program control options;
(b)   initial conditions;

(c) climatological data;
(d) time–area histogram;
(e) watershed model parameters; and
(f) historical streamflow records.

Twenty-eight parameters serve to quantify the various watershed characteristics which ultimately govern a basin's interaction with precipitation and evaporation, thereby dictating a unique hydrologic response. These parameters have been subdivided in Table 11-4 into four categories:

TABLE 11-4 Summary of parameters required for VPI and SU modification of standard watershed model[a]

Parameter	Description	Parameter	Description
a. Initial moisture conditions		c. Parameters estimated from historical record	
1. FLZS	Current lower zone soil moisture storage	1. CSSR	Streamflow channel routing parameter
2. GWS	Current groundwater slope index and gives indication of antecedent moisture conditions	2. EPXM	Maximum interception rate for a dry watershed (in./hr)
3. RI	Discharge rate at end of previous day	3. FK1	Ratio of average rainfall on the watershed to the average rainfall at recording gauge
4. SGW	Groundwater moisture storage	4. FK3	A measure of rate of loss through evapotranspiration
5. UZS	Current upper zone moisture storage	5. FKGS	Antecedent groundwater index
		6. FLIRC	Interflow recession constant
b. Parameters obtained from watershed characteristics		7. FLKK4	Groundwater recession index
1. AREA	Watershed area (acres)	d. Parameters estimated by trial and adjustment	
2. C	Time delay histogram	1. CBI	Infiltration index
3. ETL	Fraction of area in stream surface	2. CC	Interflow index
4. FIA	Fraction of area having impervious area draining directly into a stream	3. CX	Index for estimating soil surface moisture storage
5. FK24EL	Groundwater evaporation parameter. Equal to the fraction of area having water loss due to phreatophytes	4. EDF	Index for estimating soil surface moisture storage (adjust for seasonal variations)
6. LENGTH	Average length of travel for overland flow (ft)	5. EF	Evaporation–infiltration factor (adjust for seasonal variations)
7. MANI	Average Mannings roughness coefficient for impervious surfaces	6. FK24L	Indicator of water entering or leaving basin
8. MANP	Average Mannings roughness coefficient for pervious surfaces	7. FKV	Index used to provide curvilinear base flow recession
9. S	Average watershed slope (ft/ft)	8. FIZSN	Nominal lower zone storage index

[a] After Herricks and Shanholtz [2].

(a) initial moisture conditions;
(b) parameters obtained from watershed characteristics;
(c) parameters estimated from historical records; and
(d) parameters normally estimated by trial and adjustment.

The detailed criteria for estimating these parameters are given by Shanholtz *et al.* [7]. The most troublesome parameters to estimate are those listed in group (d).

Model Optimization

The procedure for determining at what point during calibration an acceptable fit has been achieved is a matter of personal judgment based upon experience. Producing a good fit was made even more difficult when data are sparse, particularly precipitation data. Precipitation is considered one of the most critical input variables.

In general, three methods have been used for achieving objectively a subjective best fit. The graphical displays have been made between simulated and recorded flow sequences; visual interpretation has been used to evaluate goodness of fit. This can be very misleading if careful attention is not given to scaling the plotted data. Correlations between estimated and recorded flows for a number of different flow sequences have also been used to guide adjustments. Although this method is somewhat objective, it gives little insight to the accuracy of quantitative estimates but does provide a good measure of relative response of one flow to another.

A Case Study

The overall objective of the study was to explore possible methods of providing a more reliable, efficient, and effective tool to evaluate potential environmental hazards from a given surface mining strategy. Such a tool, if successfully developed, could provide the decision maker with estimates of environmental consequences from which judgments could be made whether to mine and, subsequently, how to mine with minimal environmental impact on those streams draining the area.

Several studies were involved. They included

(a) generating synthetic daily streamflow data for ungauged watersheds;
(b) developing a model to predict SO_4 concentrations in the stream system; and
(c) determining particles on stream bed that can be moved as a function of stream discharge.

Only part (a) will be presented here. Parts (b) and (c) are presented in Chapter 16.

The Indian Creek Valley in Fayette County, Pennsylvania is underlain by coal deposits which have been mined since the early 1900s. Both surface and subsurface mining methods were used in the area. Although all deep mines have been shut down since 1950, some surface mining is still in progress on the watershed. The pyritic minerals exposed by mining result in highly acidic mine discharges. The effect of acid mine drainage on stream biota, and the recovery of the stream communities from this stress was the subject of a two-year study by Herricks and Cairns [8]. The watershed used in this study was that part of the Indian Creek basin located above the reservoir at Normalville, Pennsylvania. This drainage area was further divided into six subbasins: Indian Creek (upstream from Champion Run), Champion Run, Back Creek, Indian Creek (upstream from Poplar Run), Poplar Run, and Laurel Run, as shown in Fig. 11-7.

Indian Creek lies in a shallow valley in the Appalachian Plateau Province, Allegheny Mountain section of southwestern Pennsylvania. Total relief is about 1750 ft with the stream valley lying between Laurel Hill and Chestnut Ridge. The stream valley is associated with the main structural

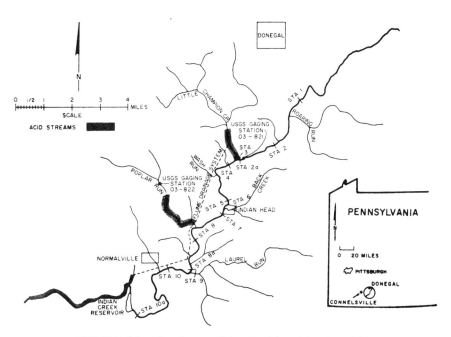

FIG. 11-7. Map of study area. (After Herrick and Shanholtz [2].)

feature of the area, the Ligonier Syncline. The rocks of the Indian Creek valley through which the main stem of Indian Creek flows are of Pennsylvania age, consisting of sandstones and shales interbedded with coal deposits. Laurel Hill and Chestnut Ridge are made up of Mississippian rocks which include the fossiliferous Greenbriar limestone and the siliceous Loyalhanna limestone. The land use was variable. The headwaters of the mainstem of Indian Creek and the tributaries were heavily forested while the valley floor was, for the most part, cleared land devoted to agricultural production.

The precipitation data were obtained from the US Weather Bureau for two stations: (a) Connelsville, located 7 miles west of Normalville, and (b) Donegal, located in the northwestern quadrant of the Indian Creek watershed. Only daily rainfall data were available for Donegal, while both hourly and daily rainfall data were available for the Connelsville Station. Continuous daily recorded discharge data were available from US Geological Survey for the Poplar Run subbasin. In addition, periodic discharge measurements were available for the Indian Creek basin upstream from Champion Run. Other data requirements, such as evaporation, land use, stream geometry, etc., were obtained from either published sources or from reconnaissance of the area.

The results of the calibration analysis are summarized in Tables 11-5 and 11-6. Table 11-5 compares two major storm sequences. These were the largest storms on record and will be referred to in Chapter 16 involving sediment movement. Table 11-6 compares recorded and simulated monthly and annual streamflows. Graphical displays of the flows for 1962–1963 are shown in Fig. 11-8. The results were accepted as sufficient to generate

TABLE 11-5 Storm hydrographs for two selected time periods[a]

Date September 1971	Simulated (cfs)	Recorded (cfs)	Date June 1972	Simulated (cfs)	Recorded (cfs)
10	0.26	1.80	20	0.14	4.80
11	2.42	2.40	21	27.15	6.80
12	33.42	37.99	22	294.11	185.99
13	64.51	100.99	23	878.75	875.94
14	883.17	867.94	24	204.65	350.48
15	170.32	139.99	25	91.52	140.99
16	166.54	121.99	26	49.36	69.99
17	110.05	89.99	27	36.45	43.00
18	58.81	48.99	28	29.61	25.00
19	68.91	45.00	29	24.55	50.00
20	61.45	43.99	30	20.53	38.00

[a] Poplar Run watershed, Fayette County, Pennsylvania. (After Herricks and Shanholtz [2].)

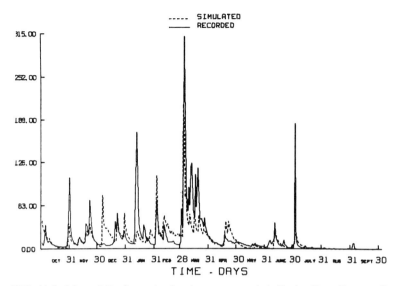

FIG. 11-8. Mean daily flows simulated versus recorded, Poplar Run, Fayette County, Pennsylvania, 1962–1963. (After Herrick and Shanholtz [2].)

flow sequences for subsequent water quality modeling efforts. This effort will be reported in Chapter 16. The final model parameters are shown in Table 11-7.

The watershed model, as calibrated on Poplar Run was used to generate 10-yr flow sequences for Indian Creek and the remaining subareas. Model parameters for these basins were evaluated based upon the modeling of Poplar Run and the experience and judgment of the investigators. The only parameters to be estimated were those which reflected the effects of varying land use, management practice, or topographic conditions.

Model Limitations

The same basic limitations which apply to any of the parametric models discussed so far also apply to the Stanford model. However, effective use of the model involves considerable practice and the reward is seemingly that a detailed solution of the flow process can be achieved.

11-4 Conclusions

Consideration of the use of the Stanford or the TVA daily flow model should be placed in the context of the trade-off diagram in Chapter 1. With 28 parameters, use of an objective best fit criterion is precluded,

TABLE 11-6 Recorded (R) and estimated (E) streamflows for Poplar Run watershed. Error (D) presented as estimated minus recorded (in.)[a]

Water year	Oct.	Nov.	Dec.	Jan.	Feb.	Mar.	Apr.	May	June	July	Aug.	Sept.	Annual
						Water yield and error summary for period							
1962–1963	1.42	1.72	2.86	1.90	3.40	5.51	1.82	0.39	0.64	0.53	0.02	0.00	20.23(E)
	1.54	2.27	1.77	3.29	1.94	9.57	1.24	0.71	0.94	1.13	0.06	0.11	24.56(R)
	-0.11	-0.55	1.10	-1.39	1.46	-4.06	0.58	-0.32	-0.29	-0.60	-0.05	-0.10	-4.34(D)
1963–1964	0.01	2.82	3.41	4.04	3.72	6.21	9.02	2.46	2.75	0.23	0.02	0.04	34.73(E)
	0.04	1.51	1.54	3.96	1.81	8.86	6.51	1.57	1.15	0.24	0.22	0.14	27.54(R)
	-0.03	1.31	1.87	0.08	1.90	-2.64	2.51	0.89	1.60	-0.01	-0.20	-0.09	7.19(D)
1964–1965	0.33	1.94	4.39	5.17	2.77	4.03	3.73	2.48	0.47	0.01	0.03	0.03	25.37(E)
	0.24	2.26	4.49	5.03	2.71	4.68	3.22	1.75	0.35	0.04	0.04	0.10	24.89(R)
	0.09	-0.32	-0.10	0.14	0.07	-0.65	0.51	0.73	0.12	-0.03	-0.01	-0.07	0.47(D)
1965–1966	0.90	1.59	2.13	4.13	4.02	2.99	4.06	2.76	0.14	0.02	0.06	0.02	22.82(E)
	0.45	1.16	0.87	2.72	5.87	2.62	3.84	2.27	0.21	0.04	0.08	0.03	20.15(R)
	0.45	0.44	1.27	1.41	-1.86	0.38	0.21	0.49	-0.07	-0.02	-0.02	-0.01	2.66(D)
1966–1967	0.09	0.75	2.44	0.64	2.18	8.24	3.55	5.04	0.39	0.06	0.06	0.03	23.46(E)
	0.10	0.83	2.16	0.97	2.05	9.72	3.29	5.23	0.37	0.24	0.10	0.20	25.23(R)
	-0.01	-0.09	0.28	-0.33	0.13	-1.48	0.27	-0.09	0.01	-0.18	-0.04	-0.17	-1.77(D)

1967–1968	0.72	2.02	2.36	2.70	1.69	3.28	1.54	6.14	1.57	0.01	0.02	0.01	22.06(E)
	0.93	1.95	1.78	2.57	1.86	3.27	1.53	5.51	1.39	0.03	0.14	0.05	21.01(R)
	−0.22	0.07	0.58	0.13	−0.16	0.01	0.00	0.63	0.17	−0.02	−0.12	−0.03	1.05(D)
1968–1969	0.01	0.95	2.50	2.13	1.69	0.81	4.06	1.94	0.55	0.87	3.40	0.18	19.08(E)
	0.05	1.98	3.75	2.70	1.69	1.54	5.26	1.74	0.39	1.02	1.73	0.38	22.24(R)
	−0.04	−1.03	−1.26	−0.56	0.00	−0.74	−1.20	0.20	0.16	−0.15	1.68	−0.21	−3.16(D)
1969–1970	0.47	2.19	5.80	5.59	3.12	3.16	3.52	2.57	2.31	0.21	0.28	0.55	29.77(E)
	0.67	2.36	3.33	4.26	3.42	4.87	4.27	2.27	1.11	1.33	0.81	0.35	29.05(R)
	−0.20	−0.17	2.46	1.32	−0.30	−1.72	−0.75	0.31	1.20	−1.12	−0.53	0.21	0.72(D)
1970–1971	1.77	1.77	4.32	3.08	4.76	3.22	0.78	3.37	3.71	1.57	2.25	7.96	38.56(E)
	0.64	1.46	4.50	3.58	6.09	3.69	1.32	3.95	1.41	0.51	2.60	6.82	36.58(R)
	1.13	0.31	−0.18	−0.50	−1.33	−0.46	−0.53	−0.58	2.30	1.05	−0.35	1.14	1.98(D)
1971–1972	0.38	1.08	2.93	3.44	5.51	4.98	4.28	1.41	6.87	1.27	0.02	0.02	32.20(E)
	0.38	1.42	3.55	3.28	2.96	8.90	5.79	2.52	7.64	3.35	0.40	0.10	40.28(R)
	0.00	−0.34	−0.62	0.16	2.54	−3.91	−1.50	−1.11	−0.76	−2.08	−0.38	−0.08	−8.08(D)

[a] After Herricks and Shanholtz [2].

TABLE 11-7 Model parameters for Poplar Run watershed[a]

Parameter	Value	Parameter	Value
FLZS	1.50	CSSR	0.90
GWS	1.00	EPXM	0.16
RI	0.0050	FK1	1.00
SGW	1.0	FK3	0.270
UZS	0.0	FKGS	0.97
		FLIRC	0.185
AREA	5932.8	FLKK4	0.836
C	0.88; 0.12		
ETL	0.001	CBI	4.411
FIA	0.0	CC	1.048
FK24EL	0.0	CX	0.540
LENGTH	800.0	EDF	0.656
MANI	0.001	EF	0.125
MANP	0.250	FK24L	-0.35
S	0.013	FKV	0.97
		FLZSN	2.00

[a] After Herricks and Shanholtz [2].

however more process detail can be included. The TVA daily flow model is objectively optimized but the level of complexity is low. Hence, a contrast in models has been presented and the choice is up to the modeler.

Problems

11-1. Contrast the use of the TVA daily flow and Stanford Watershed models in evaluating the effects of stripmining on streamflow.

11-2. Prove that the parameter B in the TVA daily flow model is a volumetric parameter used to preserve mass balance.

11-3. On how small of a watershed would you feel comfortable in using the TVA daily flow model?

11-4. Contrast the use of the TVA stormwater and daily flow models. Under what circumstances, if any, would it be appropriate to use both?

11-5. Use the TVA daily flow and the Stanford Watershed models to construct a watershed in your region. After calibration of both, project the effects of a hypothetical, but possible, land use change on the streamflow regime. Contrast the results.

11-6. Do you agree that changes in the streamflow regime of the New River Basin, Tennessee were detected? If so, to what reliability?

11-7. Do you agree that changes in the streamflow regime of the Indian Creek Valley watershed in Pennsylvania were detected? If so, to what reliability?

11-8. Formulate and test your own model for evaluating the effects of stripmining on streamflow. Contrast it with the TVA daily flow and Stanford models.

References

1. Tung, H. S., Impacts of coal mining on streamflow, a case study of the New River Watershed, Tennessee, Ph.D. Dissertation presented to the Dept. of Civil Engr., Univ. of Tennessee, Knoxville, Tennessee, Aug. 1975.
2. Herricks, E. E., and Shanholtz, V. O., Hydrologic and water quality modeling of surface water discharges from mining operations, Dept. of Agr. Engr., Virginia Poly. Inst. and State Univ., Blacksburg, Virginia, Res. Rept. No. 159, Jan. 1975.
3. Betson, R. P., A continuous daily streamflow model, TVA, Res. Paper No. 8, Knoxville, Tennessee, 1972.
4. Green, R. F., Optimization by the pattern search method. TVA, Res. Paper No. 7, Knoxville, Tennessee, 1970.
5. Viessman, W., Jr., Harbaugh, T. E., and Knapp, J. W., "Introduction to Hydrology." Intext, New York, 1972.
6. Crawford, N. H., and Linsley, R. K., Digital simulation in hydrology, Stanford Watershed Model IV, Tech. Rept. No. 39, Dept. of Civil Engr., Stanford University, California, 1966.
7. Shanholtz, V. O., Burford, J. B., and Lillard, J. H., Evaluation of a deterministic model for predicting water yields from small agricultural watersheds in Virginia, Res. Div. Bull. 73, Virginia Poly. Inst. and State Univ., Blacksburg, Virginia, 1972.
8. Herricks, E. E., and Cairns, J., Jr., Rehabilitation of streams receiving acid mine drainage, Bull. 66, Virginia Poly. Inst., Virginia Water Resources Res. Center, Blacksburg, Virginia, 1976.

Chapter 12 | Sensitivity Analysis

12-1 Need for Sensitivity Analysis

In Chapter 1, the basic principles of model sensitivity analysis were discussed. Investigation of the sensitivity of model parameters to model performance is an integral and vital part of the modeling process. Sensitivity analysis assists in answering questions concerning the relative importance of the various model components in representing the rainfall-runoff process and the accuracy needed in estimating model parameters on ungauged watersheds.

The US Geological Survey [1] model will be utilized herein as an example of sensitivity analysis. Indeed, the work of the USGS group has been a pioneering effort in the area of effective use of sensitivity analysis.

12-2 USGS Model

The USGS model [1] was deliberately kept simple in order to emphasize the parameter sensitivity and optimization aspects of the work. Shown in block flow diagram in Fig. 12-1, the model has only four storage elements with simple hydrological characteristics. The model is open to improvement as the study proceeds, viz., by making use of increased hydrological

238

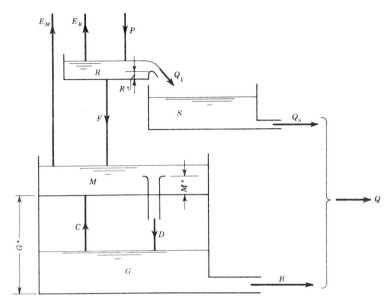

FIG. 12-1. The USGS model. (After Dawdy and O'Donnell [1].)

knowledge, by incorporating the results of component model studies, and by using more efficient optimizing procedures [1]. The most complex treatment is that which governs the infiltration component.

The components of the various elements shown in Fig. 12-1 are:

Surface storage R.   Augmented by rainfall P. Depleted by evaporation E_R, infiltration F, and, when R exceeds a threshold R^*, channel inflow Q_1 begins.

Channel storage S.   Augmented by infiltration F and capillary rise C. Depleted by transpiration E_M, and, when M exceeds a threshold M^*, deep percolation D occurs.

Ground water storage G.   Augmented by deep percolation D. Depleted by capillary rise C, and baseflow at the gauging station B. If and while G exceeds G^*, M is absorbed into G, C and D no longer operate, but E_M and F now act on G.

There are nine parameters which control the operation of the model. At the beginning of each time interval, the volume in R lies between zero and a parameter R^*. P is added to R; E_R (if any) is then called. F is calculated according to a criterion based upon a Horton-type equation which operates on the rate of supply now available from surface storage and on a potential rate of infiltration at the start of the time interval. This involves maximum and minimum infiltration rates f_0 and f_c, and an exponential die-away exponent k. These are three more parameters. As initial values for the next interval, a potential rate of infiltration f_i is calculated for the end of the current interval. Q_1 is determined to be the excess (if any) over R^* left in surface storage after E_R and F have been abstracted.

S is assumed to be a linear storage having a storage constant K_S, which is the fifth parameter. Q_S is treated as a function of the volume in S at the beginning of the time interval, of the inflow Q_1 and of K_S. A simple water budget yields the volume left in S which is ready for the beginning of the next interval.

At the beginning of each time interval, M lies between zero and the sixth parameter M^*. B_M is removed or F is added. One of these two will be zero depending on whether or not E_R satisfied E_P, the potential evapotranspiration. One of several alternatives is now followed depending on whether or not G at the start of the interval is greater than the seventh parameter G^* and, if not, whether or not the quantity in M is now greater than M^*. If G is less than G^*, D is set equal to the excess (if any) over M^* now in M; C is zero if D exists, otherwise it is determined as a function of demand in M, of supply in G, and of a maximum rate of rise c_{max} which is the eighth parameter.

M is left at M^* if D exists, otherwise it is augmented by C. At the beginning of the interval if G is greater than G^*, F, if any, acts on G directly rather than D, and C similarly in place of E_M. In this alternative, M remains at M^*.

G is assumed to be a linear storage having a storage constant K_G, which is the ninth parameter. B is then a function of the volume in G at the start of the interval, of the inflow D or abstraction C, and of K_S. Again, a water budget determines the volume left in G ready for the beginning of the next time interval.

At the beginning of the first time interval, the volumes in each of the four storage elements, the potential infiltration rate, and a set of values of the nine parameters are to be specified in order for a synthesis to be initiated. Thereafter, the computations for each time interval yield the four storage volumes and the potential infiltration rate for the beginning of the next interval. A completely general optimization would include the start-of-synthesis values of these five quantities along with the nine parameter values. However, by postulating a long dry period prior to the beginning of a synthesis, all four initial storages are set to zero and it is assumed that the beginning potential infiltration rate has recovered to the maximum value f_0. In this way, the number of parameters to be optimized was kept to nine.

The input data to the model consists of P, E_P, and Q for each of the intervals of a known record, and of estimates of the initial values of the nine parameters. The model then works through the P and E_P data and calculates a runoff volume for each time interval of the record, which in general will not agree with the known Q values. The optimization technique adjusts the initial parameter values such that the differences between the calculated and known Q values are eliminated.

12-3   Model Optimization

The USGS model [1] is optimized by the method of Rosenbrock [2] which was described in Chapter 9. It is a search technique amenable to a model such as this one whereby the partial derivatives of the objective function cannot be stated in usable analytical form.

12-4   A Case Study

A study using the USGS model is presented. The study [1] concentrated primarily upon demonstrating some principles and guidelines necessary to performing and interpreting optimization and sensitivity analysis. A second study [3] utilized a real watershed system and concentrated upon an analysis of the effects of rainfall errors on model performance.

The study [1] presented here was carried out with data artificially generated by specifying a set of values to the model parameters and rainfall input.

With the error-free input and output data, a study was made of how sensitive a model response was to changes in each of the parameters. Tests of the optimization technique [2] as regards speed and effectiveness were carried out using the error-free data. Starting from a deliberately chosen "wrong" set of parameters, the rate of progress towards the known correct set used to generate the compatible data was used to develop improvements in the optimization routine.

The objective function U was defined as the sum of the squares of the differences between the recorded and synthesized runoffs for each of the time intervals of the record. Some modifications to Rosenbrock's computer program [2] were made in order to speed up execution time. The efficiency of optimization is given by the reduction in U attained for a given number of attempts. The program keeps a running sum of the number of attempts made, those that succeeded and those that failed. It was found that the *rate* of improvement of U fell off as the number of stages increased. Eventually, changes in U between stages would become negligible even though U might not be very small. Instead of stopping at this point, a modification was made to store the latest U value, set the arbitrary step sizes back to their start-of-run values and use the latest parameter values to start a new "round" of stages. This allowed further progress to be made. A further modification was to limit the number of stages per round, since most of the progress in each round was made in the early stages. It was found that limiting the number of stages per round to six gave the most rapid overall progress.

The program terminated whenever consecutive end-of-round U values differed by less than a specified small percentage. Other minor modifications were made, some to help prevent the search from stagnating and some to rearrange test routines in order to save execution time.

The sensitivity of the response of the model to changes in each of its nine parameters was examined. This was done by computing U values for both increases and decreases of 1, 5, and 10% in each of the parameters. A wide range of sensitivity was shown. For example, the 1% changes producing the most sensitive response were nearly 100,000 times greater than the least sensitive.

Table 12-1 shows typical results of the sensitivity of model response to 1% changes in each of the parameters for a 240-step record which consisted of a sequence of 20 discrete rainfall events all ending with several intervals of no rainfall but with appreciable evaporation, and for a 60-step record with a sequence of fewer but longer and smaller rainfalls and a few short periods of evaporation. In addition, the value of G^* was lowered and that for f_c, raised in the 60-step case. These two cases provide runoff records in which the ratio of surface flow to base flow is high in the 240-step case and low in the 60-step case. Sensitivity figures in the tables are U values. The order of parameters in the table is that of decreasing sensitivity of model response.

TABLE 12-1 1% Response sensitivity using synthetic error-free data[a]

240-Step record			60-Step record		
Parameter	Value	1% Sensitivity	Parameter	Value	1% Sensitivity
K_S	10	5×10^{-4}	K_S	10	3×10^{-4}
f_c	0.2	1×10^{-4}	f_c	0.3	3×10^{-5}
f_0	2	4×10^{-5}	K_G	40	1×10^{-5}
M^*	2	4×10^{-5}	f_0	2	4×10^{-6}
k	2	2×10^{-5}	M^*	2	4×10^{-6}
K_G	40	3×10^{-6}	k	2	2×10^{-6}
R^*	0.1	1×10^{-6}	R^*	0.1	1×10^{-7}
G^*	4	2×10^{-7}	G^*	3	4×10^{-9}
c_{max}	0.1	2×10^{-7}	c_{max}	0.1	4×10^{-9}

[a] After Dawdy and O'Donnell [1].

To free the optimization studies from errors in any real data records, the model was first used to generate a synthetic set of runoff values from a set of arbitrary parameter values and a made-up record of P and R_p values. Not only does this provide error-free data, but we now know the "correct" parameter values towards which any other set of values should be optimized.

Most of the development of the optimization aspect of the overall model was done with a 60-step record of synthetic data. Table 12-2 gives the results of a test run with these data in which five complete rounds with each of the six stages plus a final incomplete round of the five stages were made. The starting values of the parameters were, with one exception, 50% above or below their correct values.

Seven of the nine parameters were optimized to within 15% of their correct values and five were within 3% while two are still a long way off their correct values.

A comparison made between Tables 12-1 and 12-2 led to the conclusion that the greater the sensitivity of the model response to a parameter, the closer and sooner that parameter will be optimized. Parameters for which the response sensitivity is two or more orders of magnitude less than that of the other parameters optimized poorly.

12-5 Conclusions

Each refinement to a model adds more parameters with a marginally diminishing net gain in accuracy. The added parameters will be less sensitive than the original ones. Yet, even though the model will appear more realistic, the fitted parameters will reflect less and less in their numerical values. This is a natural consequence of moving to the right along the abscissa scale in the trade-off diagram in Chapter 1, Fig. 1-2. Further, small errors in the hydrologic data may generate large errors in some of the less sensitive parameters. Apparently the residual errors in the computed values may be small, notwithstanding that some of the more insensitive parameters have large errors in them.

It should be recognized that the goodness of fit of the model to the observed data, in the components as well as in any overall model, is a function of the structure of the mathematical model, the accuracy of the data used, the method of fitting the model to the data, and the criteria used for goodness of fit, i.e., the objective function. The first two are somewhat related to the closeness of fit, but the last two may be independently important. Any fitting method makes assumptions about the model fitted primarily because the two must be mathematically compatible. As pointed out in

TABLE 12-2. Results of parameter optimization on error-free data[a]

Parameter	Correct value	Starting value	Parameter value at end of round				Residual difference (%)
			1	2	4	6	
K_S	10	15	10.17	10.13	10.011	10.015	0.15
f_c	0.2	0.1	0.1721	0.1700	0.1973	0.1972	1.4
k	2	3	2.931	2.113	1.983	1.970	1.5
f_0	2	1	1.952	1.943	1.972	1.967	1.7
M^*	2	1	1.815	1.886	1.936	1.947	2.7
K_G	40	35	31.32	57.10	45.17	43.96	9.9
R^*	0.1	0.15	0.8059	0.2615	0.1174	0.1143	14
G^*	4	6	5.834	18.03	19.82	19.27	380
c_{max}	0.1	0.15	2.049	0.6363	0.5282	0.5574	460
U value	5.07×10^{-1}		9.04×10^{-4}	1.23×10^{-4}	5.43×10^{-6}	2.91×10^{-6}	

[a] After Dawdy and O'Donnell [1].

earlier chapters, the shape of the response surface will determine to a great extent the viability of the optimization. Therefore, the last two criteria mentioned above are interrelated, because the response surface is the mathematical statement of the test of the goodness of fit.

Problems

12-1. Why is sensitivity analysis of the effect of model parameters on the objective function necessary?

12-2. There are two fundamental uses of sensitivity analysis. One is involved in model optimization and one in model verification. Contrast the two.

12-3. What are the ramifications of the objective function being very insensitive to a model parameter? Very sensitive?

12-4. How does the shape of the response surface affect optimization, and how is sensitivity analysis used to illustrate the phenomenon?

12-5. What factors affect the goodness of fit of the stormwater model to the data?

12-6. Discuss the effect of continuing refinement to a stormwater model by adding more parameters.

12-7. Discuss the need for the mathematical structure of a stormwater model to be compatible with the objective best fit method.

12-8. Apply the US Geological Survey to a watershed in your region for which there are appropriate hydrologic data available. Explore its sensitivity to input data and parameters and discuss the results in terms of the adequacy of the input data and the interaction between the model parameters.

References

1. Dawdy, D. R., and O'Donnell, T., Mathematical models of catchment behavior, *J. Hydr. Div., Amer. Soc. Civil Engr.* **91** (HY4), 123–139 (July 1965).
2. Rosenbrock, H. H., An automatic method of finding the greatest or least value of a function, *The Computer J.* **3** (1960).
3. Dawdy, D. R., and Bergman, J. M. Effect of rainfall variability on streamflow simulation, *Water Resources Res.* **5,** 958–966 (1969).

Chapter 13 | Regionalization of Model Parameters

13-1 Introduction

To regionalize a model simply means to develop a scientific basis for predicting the model parameters on ungauged watersheds from hydrologic and physiographic characteristics of that watershed. Regionalization can be accomplished if there are enough bench mark watersheds with adequate storm rainfall and runoff data such that a statistical inference may be drawn.

Few examples of model regionalization have been reported in the literature. As discussed in Chapter 9, the efforts of Wong [1] and Thomas and Benson [2] resulted in regionalized prediction equations for mean annual flood in New England and several other statistical flow properties in a number of river basins nationwide. These efforts, although they do not deal directly with stormwater, serve as good analogies to the regionalization scheme developed by Ardis [3] for the TVA stormwater model discussed in Chapter 10, which will be presented in this chapter.

13-2 Regionalization of TVA Stormwater Model

Using stepwise multiple regression, the pooled data equation for QMAX/AREA is

246

$$\frac{QMAX}{AREA} = 197 \frac{SRO^{1.13}}{NPE^{-0.57}} \tag{13-1}$$

which explained 78% of the variance. By allowing the stepwise selection of watershed characteristics to explain further variance due to among-watershed differences, the prediction equation becomes

$$\frac{QMAX}{AREA} = 197 \frac{SRO^{1.07}}{NPE^{0.47}SINU^{0.38}TIME^{0.31}SOILS^{0.09}W^{0.23}} \tag{13-2}$$

This improved the prediction equation by explaining an additional 9% of the variance. SINU is a measure of sinuosity, TIME the length divided by the square root of the slope of the mainstream, SOILS the discharge that is equal or exceeded 70% of the time, and W the percent of the watershed in forest cover.

Similarly, the peak of the unit graph was regionalized as

$$UP = 650 \frac{SHAPE^{0.28}SRO^{0.13}}{SINU^{0.85}NPE^{0.23}TIME^{0.71}DD^{3.1}W^{0.53}} \tag{13-3}$$

where SHAPE is length of the main stream divided by surface drainage area, and DD is drainage density of the watershed. Equation (13-3) explained 73% of the variance.

For time to peak of the unit graph, $T1$,

$$T1 = 0.0025 \frac{NPE^{0.19}SINU^{1.33}DD^{4.0}TIME^{0.46}}{SRO^{0.11}SHAPE^{1.16}SOILS^{0.23}} \tag{13-4}$$

Equation (13-4) explained only 58% of the variance.

The prediction equation for the location parameter $T2$ is

$$T2 = 0.0019 \frac{AREA^{0.40}SINU^{1.0}DD^{3.7}NPE^{0.14}}{S_c^{0.31}SHAPE^{0.26}SOILS^{0.11}} \tag{13-5}$$

and Eq. (13-5) explained 63% of the variance. S_c is slope of the main stream.

The time base of the unit graph, $T3$, is predicted by

$$T3 = 0.285 \frac{AREA^{0.68}SINU^{0.84}DD^{2.02}}{SHAPE^{0.59}SOILS^{0.076}} \tag{13-6}$$

and Eq. (13-6) explained 75% of the variance.

And finally, the parameter prediction equation for PHI, the constant loss parameter is

$$PHI = 0.34 + 0.09 * LOSS - 0.044 * WS$$

$$- 0.0054 * NPE - 0.017 * SOILS \tag{13-7}$$

TABLE 13-1 Final parameter-prediction models (system of equations)[a]

Individual equations	Leading coefficient	Powers on watershed characteristics for WF1, WF2, and WF3*								Correlation coefficient
		AREA	S_c	SHAPE	W	DD	SINU	SOILS	TIME	
QMAX/AREA										
WF1	0.197				−0.227		−0.375	−0.0905	−0.308	0.955
WF2	0.364				0.261			−0.145		
WF3	−0.0617	0.0504			0.0293	0.750				
UP										
WF1	0.650			0.280	−0.530	−3.09	−0.850		−0.711	0.858
WF2	0.357				0.238			−0.142		
WF3	−0.0490	0.234			0.0189	0.341				
T1										
WF1	0.0000167			−1.24	0.582	4.79	0.816	−0.350	0.681	0.831
WF2	0									
WF3	0.0115			0.383	0.344		−0.401		0.141	
T2										
WF1	0.00188	0.396	−0.314	−0.259	0.290	3.71	1.00	−0.106		0.792
WF2	0									
WF3	0.136									
T3										
WF1	0.285	0.679		−0.558		2.02	0.843	−0.0757		0.850
WF2	0									
WF3	0									0.819

PHI = 0.0340 + 0.0903 * LOSS − 0.0445 * WS − 0.00538 * NPE − 0.0172 * SOILS

[a] After Ardis [3].

where LOSS is the difference between accumulated rainfall ARF and SRO. Equation (13-6) explained 68% of the variance.

Table 13-1 summarizes the parameter prediction equations. Other than PHI, each follows the form of Eq. (13-1). The multiple-correlation coefficients for QMAX/AREA, *UP*, and *T*1 are from optimization with PATSEAR. The remaining are from stepwise multiple regression.

Nine watershed characteristics were available for selection by stepwise multiple regression to explain among-watershed response variation. However, several are general measures of similar characteristics and only one was allowed to be considered in each analysis. For example, S_c and *SL* (weighted mean slope of the watershed) are both measures of slope, but each analysis for a given response parameter would allow selection consideration of only one at a time; i.e., one analysis would allow S_c to be selected while *SL* was excluded and vice versa in the next analysis. A similar procedure was followed with AREA and TIME since both are relative measures of size. Hence, in effect, only seven independent watershed characteristics were available for selection consideration by stepwise multiple regression to explain among-watershed response variation in any given analysis. Since *SL* was not selected as being significant in any of the final parameter-prediction models, the effective number of watershed characteristics used in the study was eight. Table 13-2 lists the eight watershed characteristics used in Table 13-1. This effort is an attempt to keep interrelationships out of the parameter prediction equations. However, it would be interesting to have applied principal components analysis to the development of the prediction equations.

TABLE 13-2   Study watershed characteristics[a]

Study watersheds	AREA	S_c	SHAPE	*W*	*DD*	SINU	SOILS	TIME
1. Crab Creek	10.9	71.7	1.66	82.9	12.26	0.16	1.59	0.58
2. Boylston Creek	14.8	22.8	6.63	64.3	13.03	0.17	1.35	2.43
3. S. F. Mills River	9.99	54.5	1.41	100	9.54	0.33	1.53	0.68
4. Allen Creek	14.4	297	1.49	98.7	11.64	0.11	1.12	0.30
5. North Indian Creek	15.9	151	2.31	78.7	12.76	0.18	0.54	0.58
6. Noland Creek	13.8	331	3.78	99.9	12.86	0.18	1.47	0.47
7. North Fork Citico Creek	7.04	394	2.69	100	14.29	0.24	1.23	0.27
8. Little Chestuee Creek	8.24	25.2	2.34	49.4	12.37	0.13	0.57	0.99
9. Cane Creek	16.8	14.0	2.26	23.4	12.44	0.13	0.26	1.86
10. Mill Creek	7.18	31.0	3.41	41.5	10.97	0.20	0.058	1.07
11. Bear Creek	13.9	24.2	1.75	85.5	10.69	0.26	0.23	1.26

[a] After Ardis [3].

The relatively low correlation coefficients resulting from the parameter prediction equations may have been attributed to such a small data base covering such a large geographical area. However, the data base is still one of the most extensive available.

13-3   Example Applications

The parameter prediction equations developed regionalized the six parameters associated with the stormwater response of the study of watersheds in the Tennessee Valley. In each case, the definitions of a given response parameter involves watershed characteristics; in most, storm characteristics were also a part of the definition. The watershed characteristics found significant during development of the models included some measures that are physically fixed and others that can be manipulated by man.

EXAMPLE 13.1   As a test of the validity of the regionalization scheme, two very different watersheds, not used in developing the models, were set aside to predict stormwater hydrographs. Beech River, on the western edge of the Tennessee River Basin in the Mississippi Alluvial Plain physiographic province, is a flat agricultural watershed. Catheys Creek is a very steep forested watershed in the Blue Ridge physiographic province on the extreme eastern edge of the Tennessee River Basin. The watershed characteristics of each as well as others that will be mentioned subsequently are listed in Table 13-3.

Eleven storms were simulated on Beech River, and the results are summarized in Table 13-4. Peak coordinates for the simulated Beech River storm hydrographs, using the double-triangle model, compared well with the observed storm hydrographs. The simulated peak discharge had an average absolute error of 25% and ranged from $+26$ to -57%. The average simulated time to peak was, on the average, one hour less than the observed

TABLE 13-3   Test watershed characteristics[a]

Test watersheds	AREA	S_c	SHAPE	W	DD	SINU	SOILS	TIME
A. Catheys Creek	11.4	149	2.22	100	14.53	0.17	1.88	0.48
B. Beech River	15.9	8.2	1.97	34.3	13.87	0.14	0.64	2.22
C. North Potato Creek	13.0	58.1	3.50	30.7	15.78	0.19	1.41	1.05
D. Chestuee Creek	14.8	13.3	3.62	17.5	9.87	0.15	0.42	2.30
E. Little Bear Creek	36.2	10.9	3.19	62.0	13.91	0.52	0.35	4.92
F. Whitehead Creek	1.92	49.6	2.11	29.0	11.29	0.24	1.10	0.35

[a] After Ardis [3].

TABLE 13-4 Simulation tests[a]

Watershed	Storm date		Peak discharge			Time to peak		r^2
			Obs. (cfs)	QMAX (cfs)	Sim. (cfs)	Obs. (hr)	Sim. (hr)	
Beech River	January	20, 1954	1240	1070	811	18	18	0.856
	January	29, 1956	1660	1310	1030	14	15	0.726
	February	27, 1962	1060	584	451	18	20	0.514
	March	11, 1963	672	550	483	19	19	0.914
	September	30, 1955	766	1112	759	15	12	0.774
	March	6, 1961	452	492	420	22	20	0.895
	July	22, 1959	1150	1670	1010	10	10	0.959
	March	26, 1962	586	1180	692	11	8	0.755
	April	11, 1962	1420	1300	955	16	16	0.866
	August	30, 1960	386	714	485	16	11	0.511
	April	4, 1957	1260	1540	1030	11	11	0.937
	Average		968	1047	739	15.5	14.5	0.792
Catheys Creek	August	26, 1949	416	383	354	18	20	0.819
	March	11, 1952	1170	727	672	19	21	0.694
	March	22, 1952	800	502	431	17	21	0.663
	January	22, 1954	309	384	353	16	19	0.806
	Average		674	499	453	17.5	20.2	0.746

[a] After Ardis [3].

with an average absolute error of 10%. The average coefficient of determination from simulation was 0.792 and varied from 0.511 to 0.960 with a mean of 0.856. The QMAX/AREA model did not simulate the peak discharge for the 11 events as well; the average absolute error was 38%. Figure 13-1 compares the best and poorest simulations for Beech River.

The average coefficient of determination from simulation for the four events on Catheys Creek was 0.746 with a range of 0.663 to 0.819. Peak coordinates of the simulated storm hydrographs were late (average absolute error of 16%) and low (average absolute error of 29%). The QMAX/AREA model also simulated low; the average absolute error was 27%. Figure 13-2 compares the best and poorest simulations for Catheys Creek.

Rainfall data were questionable on some of the Catheys Creek storms. For example, two simulations on Catheys Creek were not done because the bulk of the storm rainfall occurred after the flood peak. For those four simulated storms, the consistently delayed peak indicates probable rainfall timing problems. Hence, rather than say that the model did not simulate

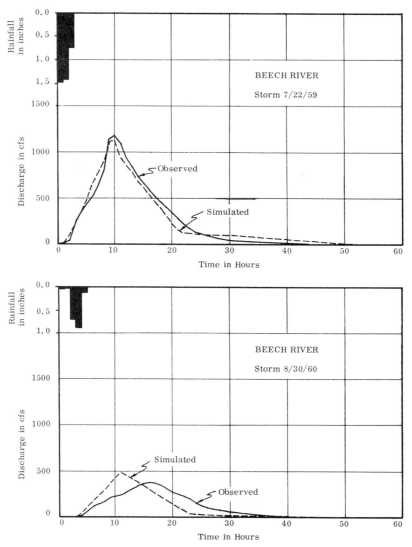

FIG. 13-1. Best and poorest simulation versus observed hydrographs for two storms at Beech River test watershed. (After Ardis [3].)

Catheys Creek too well, it can be said that the rainfall time distribution is not well known. Obviously, good simulations cannot be made with inadequate rainfall data. Rainfall data for the Beech River watershed were not questionable.

It would have been desirable to have more watersheds for simulation

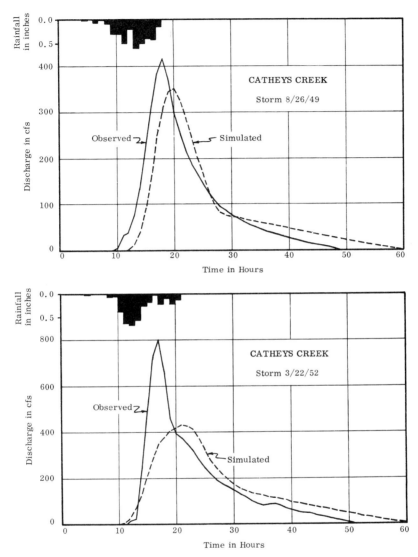

FIG. 13-2. Best and poorest simulation versus observed hydrographs for two storms at Catheys Creek test watershed. (After Ardis [3].)

tests, but it was necessary to use as many as possible from the total sample for model calibration. Beech River and Catheys Creek were set aside for testing because they are extremely different. In general, the simulation test results for Beech River and Catheys Creek, as summarized in Table 13-4, were considered good. Those for Beech River are better.

Four other watersheds were not included in the set used for calibrating the parameter prediction equations. Two of these were not included because of unusual watershed characteristics not involved in the 11 study watersheds. They are North Potato Creek and Chestuee Creek. As seen in Table 13-3, both are in the size range selected for the study. The remaining watershed characteristics are fairly well in line with those from the study watersheds.

EXAMPLE 13.2   North Potato Creek is a 13.0-sq mile watershed with 30.7% of its area in forest. The remaining portion of the drainage area is completely denuded. The forested area lies in the headwaters of this watershed, which at one time was entirely in forest. Copper mining in the area, begun in 1850, produced large volumes of sulfur dioxide-laden smoke which had killed all vegetation in the lower two-thirds of this watershed by the turn of the century. The denuded area is extremely gullied and runoff is very rapid. The percent of area in forest W, in this case 30.7%, is the only measure of land use identified for use in this research. It is in no way a measure of that area not in forest. In general, however, for the study region, those lands not in forest are generally used for agriculture. The parameter prediction models are based upon this assumption. The invalidity of this assumption for the North Potato Creek watershed was tested on three storms. A typical example is shown in Fig. 13-3. If the denuded basin had

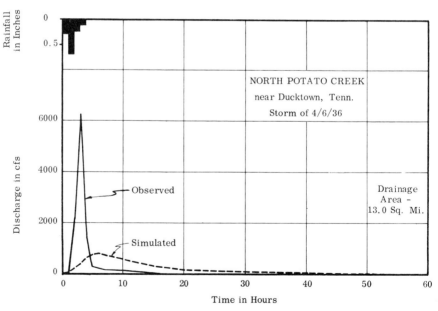

FIG. 13-3. Reduction in streamflow due to conversion of a denuded watershed to agriculture and pasture. (After Ardis [3].)

been in agriculture, the resultant storm hydrograph would be much like that shown as simulated. Rather than reaching the observed peak of 6230 cfs in 3 hr, it would have amounted to only 802 cfs peaking in 6 hr. If the denuded area could be converted to agricultural use, gully erosion would be eliminated and flooding drastically reduced.

EXAMPLE 13.3   Chestuee Creek watershed is a 14.8-sq mile agricultural watershed with extensive semi-karst topography. Sinkholes apparent on topographic maps influence the drainage collection of rainfall for nearly 5% of the total drainage area. The extent of underground solution channels in the limestone of the area is not known. Caves and caverns are common. How karst topography quantitatively affects surface–water hydrology is not known at this time. An indirect approach to such an answer was used via simulation. Ten storm hydrographs and the associated rainfall were abstracted for test purposes. Other than the largest flood of record (QMAX = 4050 cfs), most simulations significantly oversimulated the peak discharge and indicated that it should occur nearly 4 hr sooner than the average time to peak of 20 hr (20%). A typical simulation is shown in Fig. 13-4. Notice that the influence of the karst topography is to delay the entire response and allow for longer contributions to runoff from the watershed. The peak discharge is about one-half of what it would have been if the area

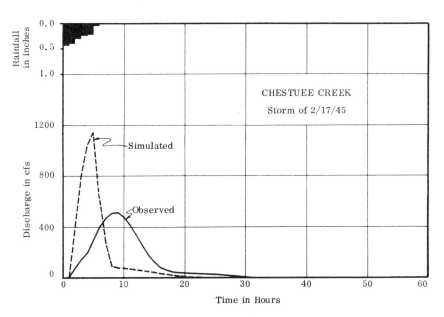

FIG. 13-4. Simulated hydrograph illustrating the effect of semi-karst topography which is not characterized by the model. (After Ardis [3].)

were not karst. The above results are rational from a qualitative standpoint; they are now quantified.

13-4   Limitations of Regionalized Models

The reliability of a regionalized model can always be improved by incorporating a larger data base into the analysis. The data base analyzed in the scheme presented in this chapter is principally from agricultural watersheds. Hence, the effects of urbanization, logging, and strip mining are not included. More hydrologic data are needed to expand the reliable domain of application of the regionalized scheme. However, consulting engineering will continue notwithstanding these gaps in our knowledge. Hence, utilization of this scheme, at least in the Tennessee Valley, still constitutes the most scientifically based prediction scheme. The effective utilization of this scheme will depend largely upon the knowledge and skill of the modeler regarding the stormwater regionalization scheme.

Problems

13-1.   What does regionalization of parameters mean, and what is its utility?

13-2.   Test apply the TVA stormwater model by predicting storm hydrographs on a gauged small watershed in your region. What do you conclude?

13-3.   How could sensitivity analysis be useful in improving the utility of the regionalization equations associated with the TVA stormwater model?

13-4.   Predict the change in the peak stormwater flow for a given rain associated with development in the Eighth Creek watershed in Knoxville, Tennessee. The design storm has 2 in. of rain occurring in 30 min and the CN before and after development are 65 and 90, respectively. The parameters for the watershed before development were: AREA = 2.7 sq miles; $S_c = 25.2$; SHAPE = 2.61; $W = 50.9$; $DD = 12.6$; SINU = 0.40; SOILS = 1.51; TIME = 3.62. After development, all parameters remained essentially constant except $W = 15.1$; SOILS = 5.23. Use a unit-response function associated with a 5 min rainfall duration and assume that there would be no baseflow.

13-5.   Use the TVA stormwater model to predict the effects of changes in stormflow associated with a real or anticipated land use change in your region and for a given design rainstorm. Discuss the reliability of the prediction and properly qualify it.

13-6.   How can regionalization schemes be improved?

13-7. Can we ever have enough hydrologic data? Could there be an optimum amount, quality, and type? Discuss.

13-8. What does effective utilization of a regionalization scheme depend upon?

13-9. How can regionalization of stormwater models be developed into a science?

References

1. Wong, S. T., A multivariate statistical model for predicting mean annual flood in New England, *Ann. Amer. Assoc. Geographers* **53,** No. 3, 298–311 (Sept. 1963).

2. Thomas, D. M., and Benson, M. A., Generalization of streamflow characteristics, US Geological Survey, Open-file Report, Washington, D.C., 1969.

3. Ardis, C. V., Jr., Storm hydrographs using a double-triangle model, TVA., Div. of Water Control Planning, Knoxville, Tennessee, Jan. 1973.

PART IV | Stochastic Stormwater Modeling

Chapter 14 | Stormwater Frequency Modeling

14-1 Introduction

Stochastic stormwater modeling has three basic subheadings: (1) regression analysis; (2) time series analysis; and (3) flow frequency analysis. A common characteristic of all three approaches is that there is a recognized element of uncertainty in the process being modeled. The three approaches simply treat this element of uncertainty in different manners.

Regression analysis has been discussed previously in Chapter 9 because it is in itself an optimization procedure. Hence, if the model structure is linear or of a relatively simple nonlinear form, least squares regression would be compatible with the model and serve as the optimization procedure. However, if the model is a "black box," then the regression would definitely take on a stochastic flavor. Time series analysis will not be discussed in this book simply because an entire volume would be needed to adequately cover the subject. Principally, time series analysis analyzes a continuous time series of runoff and draws an inference as to the underlying generating mechanism. Since stormwater is treated herein as occurring usually on small watersheds which do not have a significant baseflow component, time series analysis would not be an appropriate tool for analysis.

Frequency analysis is a valuable set of techniques for developing design flows where there is a continuous streamflow record available but no adequate rainfall record available. Primarily, frequency analysis is applied to runoff records for the purpose of drawing an inference as to the extremes in flow, both high and low. There are no cause and effect concepts built into this modeling effort. Flowrate is assumed to be entirely probabilistic, i.e., it is considered to be a random variable with a certain probability associated with the occurrence of that flowrate. Further, it is assumed that the underlying processes are stationary, i.e., that the statistical characteristics of the basin remain constant in time. This means simply that the response characteristics and land use must remain constant during the sampling period.

14-2   Return Period

Return period of a stormwater flow is defined as the time period for which *on the average* a stormwater flow will be equaled or exceeded. For example, if for a hypothetical watershed 2000 cfs is determined to have a 100-yr return period, this means that on the average the stream will equal or exceed 2000 cfs once every 100 yr. In flow frequency analysis each stormwater event is considered to be independent and mutually exclusive, meaning that there are no conditional probabilities. The systems are therefore treated as being without memory.

EXAMPLE 14.1   What is the probability that a peak stormwater flow with a 100-yr return period will be equaled or exceeded in any given year?

Since there are no conditional probabilities, the probability is simply

$$P = (1/100) \times 100\% = 1\%$$

Hence, in general, the probability that an event with a return period of T-years will be equaled or exceeded in any given year is

$$P(T) = (1/T) \times 100\% \tag{14-1}$$

EXAMPLE 14.2   To further illustrate the concept of return period, consider a hypothetical watershed which has exactly 100 yr of continuous streamflow records and no rainfall records. The watershed is assumed to have been in substantially the same land use conditions during the sampling period. The streamflow record is shown in Fig. 14-1.

Since Q_a is the largest flow of record, the return period model indicates that it has a 100-yr return period. Hence, return periods for the remainder of the events can be determined by searching through the record and re-

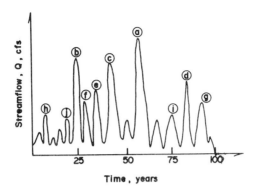

FIG. 14-1. One hundred year continuous streamflow record on a hypothetical watershed.

cording the maximum flow in each year and ranking them as shown in Table 14-1.

Q_b has a return period of 50 yr because on the average it was equaled or exceeded twice in the 100 yr period. Q_c has a return period of 33.3 yr because on the average it was equaled or exceeded three times in the 100 yr period. The return periods for the remainder of the peak stormwater events can be calculated using this procedure.

There are severe limitations to this return period model. Seldom is there exactly 100 yr of streamflow record; for that matter there are very few 100 yr records at all. Of the very few records which are 100 yr or longer, land use has changed considerably and consistently. Further, the event tagged

TABLE 14-1   Determination of return periods for record in Fig. 14-1

Event	Rank	Period (yr)	Simple probability (%)
Q_a	1	100	1
Q_b	2	50	2
Q_c	3	33.3	3
Q_d	4	25	4
Q_e	5	20	5
Q_f	6	16.7	6
Q_g	7	14.3	7
Q_h	8	12.5	8
Q_i	9	11.1	9
Q_j	10	10	10

as having a 100 yr return period may have a "true" return period of 100 yr or 50 yr. This depends upon whether the 100 yr record was in a period of wetness, drought, or somewhere in between.

Although Example 14.2 serves simply as an example illustration of return period, it is too idealistic for use in the real world. Hence, the problem of return period definition reverts to one of analyzing a streamflow record considerably shorter than 100 yr and drawing a statistical inference or extrapolation as to the higher return periods.

14-3   Probability Density Functions

The solution of the problem of stormwater frequency analysis is attained by selecting a proper probability density function, a proper plotting position for the stormwater data, and a proper criteria for fitting the density function to the data. (It will be assumed that the reader is aware of what probability density functions and cumulative distributions are. If not, see a basic engineering statistics book or Viessman et al. [1]).

There are several density functions which have been used in water resources engineering, and they have been catalogued by Viessman et al. [1] as shown in Table 14-2. The log normal, gamma, and extreme value have been

TABLE 14-2   Table of common density functions

Distribution of random variable X	Probability density function
Normal	$f(x) = \dfrac{1}{\sigma\sqrt{2\pi}} \exp[-(x - \mu)^2/2\sigma^2]$
Log normal ($y = \ln x$)	$f(y) = \dfrac{1}{\sigma_y\sqrt{2\pi}} \exp[-(y - \mu_y)^2/2\sigma_y^2]$
Binomial	$P(x) = \dfrac{n!}{x!(n - x)!} p^x(1 - p)^{n-x}$
Poisson	$P(x) = \dfrac{\lambda^x e^{-\lambda}}{x!}$
Uniform	$f(x) = \dfrac{1}{b - a}$
Exponential	$f(x) = \dfrac{1}{a} e^{-x/a}$
Gamma	$f(x) = \dfrac{x^\alpha e^{-x/\beta}}{\beta^{\alpha+1}\Gamma(\alpha + 1)}$
Extreme value	$f(x) = \alpha \exp\{\alpha(-\alpha(x - u) - \exp[-\alpha(x - u)]\}$

widely used but more recently the log gamma (now shown) has become in vogue principally because it was recommended for standard use by the US Water Resources Council [2]. The log density functions are merely transforms of the parent functions made to induce skewness or added skewness.

The choice of a density function has been traditionally made on the basis of the knowledge of the underlying process and the tractability of the function for calculating moments of the function. Matching the first two moments of the data with the first two moments of the distribution has been a traditional method for achieving best fit of the function to the data. Hence, there has been a significant element of trial and error involved in flow frequency analysis.

The best fit procedure has been with the cumulative distribution rather than with the probability density function. There are two main reasons for this development. First, the shape of the probability density function derived from the data is very sensitive to class size, and second, the mathematics associated with fitting a selected theoretical density function to the one developed with the data has been simply intractable. However, this last problem seems to have a solution as will be shown in Section 14-5, but the sensitivity due to class size needs further research.

To aid in the problems associated with fitting cumulative distributions to data, special graphs have been prepared by various investigators such that the particular cumulative distribution will plot on the graph as a straight line. Examples of use of the paper will follow.

14-4   Plotting Positions

Plotting position refers to the probability or return period that is to be assigned to each of the observed events. This provides a coordinate for plotting each of the observed events on special probability paper. Since the records analyzed will not be long, choice of plotting position has been influenced to some extent by its ability to stretch the record out. Otherwise, the highest return period for a 25-yr record will be 25 yr, or for a 10-yr record it will be 10 yr. It is recognized that there is a significant chance that a 10-yr record contains an event with a higher return period than 10 yr, etc.

Plotting positions which have been used in water resources engineering are shown in Table 14-3.

EXAMPLE 14.3   Using the plotting positions in Table 14-3, calculate the probability that an event of rank 1 in a 20-yr record will be equaled or exceeded in any given year as well as the return period.

As shown in Table 14-4 the Hazen method stretches the record out the most and the California method the least. In Table 14-1, the California

TABLE 14-3   Formulas for plotting positions

| | | $m = 1$ and $n = 10$ | |
| | | | |
Method	Position	P	T
Beard	$1 - (0.5)^{1/n}$	0.067	14.9
Blom	$(m - 3/8)/(n + 1/4)$	0.061	16.4
California	m/n	0.10	10
Chegadayev	$(m - 0.3)/(n + 0.4)$	0.067	14.9
Hazen	$(2m - 1)/2n$	0.05	20
Tukey	$(3m - 1)/(3n + 1)$	0.065	15.5
Weibull	$m/(n + 1)$	0.091	11

TABLE 14-4

Method	$P\ (\%)$	$T\ (\text{yr})$
California	5.0	20
Hazen	3.5	40
Beard	3.4	29.4
Weibull	4.8	21
Chegadayev	3.4	29.1
Blom	2.5	32.4
Tukey	3.3	30.5

method was used. Also, the Weibull plotting position appears to be the most widely used method.

14-5   Best Fit Criteria

The most widely used objective criterion for fitting theoretical cumulative distributions to data is the method of moments. The first two moments (mean and standard deviation) are calculated from the data, and are set equal to the first two moments of the selected cumulative distribution. This defines the parameters in the distributions (Table 14-2) and permits ready construction of a straight line on the special probability paper. At this point, the straight line can be extended beyond the data and projections of extreme events made.

Frequency Factors

To facilitate a ready determination of the line of best fit, Chow [3] proposed the use of the general expression

$$Q = \bar{Q} + Ks \tag{14-2}$$

where K is the frequency factor and s the standard deviation of the data. The frequency factor is the number of standard deviations above or below the mean to attain the probability of the flow of interest. For two parameter distributions (e.g., normal or log normal) K varies only with probability or return period. In skewed distributions, it varies with the coefficient of skewness and the length of record. For the log normal distribution, the frequency factor corresponds to the standard normal deviate found in readily available tables, such as those given in Appendix I. The tables of frequency factors for additional distributions are also available. By way of example, the log normal will be used in this text.

EXAMPLE 14.4   There are 25 yr of continuous streamflow records available in the First Creek watershed at Mineral Springs Avenue, Knoxville, Tennessee. The drainage area is 12.5 sq miles and the rainfall data are inadequate to apply the TVA stormwater model. The basin land use has remained substantially constant during these 25 yr.

Water Resources Council Method

Using the Weibull plotting position, the log normal distribution and the method of moments, estimate the magnitude of the flows with 100, 50, 33.3, and 25 yr return periods from the data given in Table 14-5.

The data are plotted in Fig. 14-2 on specially constructed log normal graph paper. Now, using the method of moments and Appendix I, we find the line of best fit by locating two points on the graph. The mean and standard deviation of the transformed variate are

$$\bar{y} \text{ (mean)} = 2.770, \qquad S_y \text{ (std dev)} = 0.237$$

The transformed discharge associated with a plotting position is

$$y = zS_y + \bar{y} \qquad\qquad (14\text{-}3)$$

The plotting position associated with the mean is $z = 0$, 50% greater than or equal to, or $Q = 589$ cfs. A second position chosen was $z = 1.282$ or 10% greater than or equal to, which corresponds to $y = 3.07$ or $Q = 1185$ cfs. The line of best fit can now be drawn and interpolations and extrapolations can be made. The flows associated with 100, 50, 33.3, and 25 yr return periods are picked off of the graph as shown in Table 14-6.

Nonlinear Least Squares

Recently, Snyder and Wallace [4] have reported the application of nonlinear least squares, discussed in Chapter 9, to fitting probability density functions to data. This provides an alternative procedure to fitting in the cumulative domain. There are several advantages of this approach: choice

TABLE 14-5

Year	Rank, m	Plotting position ($\% \geq$)	Max. annual flood flow, (cfs)
1957	1	3.8	1455
1963	2	7.7	1350
1969	3	11.5	1320
1948	4	15.4	1268
1951	5	19.2	890
1967	6	23.1	855
1965	7	26.9	853
1947	8	30.8	775
1950	9	34.6	755
1949	10	38.5	718
1946	11	42.3	670
1961	12	46.2	660
1956	13	50.0	635
1954	14	53.8	600
1970	15	57.7	570
1962	16	61.5	552
1952	17	65.4	420
1953	18	69.2	375
1964	19	73.1	368
1959	20	76.9	365
1966	21	80.8	347
1958	22	84.6	330
1955	23	88.5	270
1960	24	92.3	265
1968	25	96.1	224

TABLE 14-6

Return period	Design flow (cfs)
100	2080
50	1750
33.3	1600
25	1500

of a plotting position is not involved, the bias and unreliability commonly known about the method of moments is bypassed, there is no problem with outliers (to be discussed below), and many density functions may be tested efficiently on the computer as to their capability to fit the data.

An objective function is set up to minimize the sum of the squares of the observed frequencies and the fitted model frequencies as shown in Fig. 14-3.

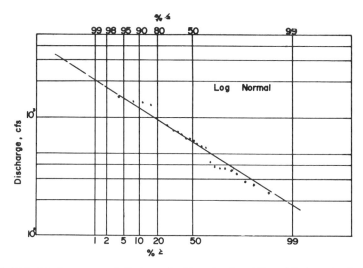

FIG. 14-2. Maximum annual discharge for First Creek at Mineral Springs Av. in Knoxville, Tennessee using log normal distribution and method of moments fitting.

The frequency data are entered into the computer program and the optimized parameters producing the best fit are calculated.

Another advantage of this method is that it is not limited to maximum monthly or even maximum daily flows. This provides great flexibility because often several large floods or stormwater flows will occur in a single month. Hence, with this method, at least twelve times as much information can be analyzed. The one major uncertainty associated with fitting in the

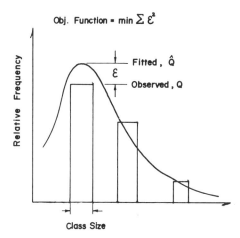

FIG. 14-3. Objective function in using nonlinear least squares to fit in density domain.

density domain rather than the cumulative distribution domain is the sensitivity of the fit to class size. However, experience has shown that the proper class size will result from experimenting with the data until the observed frequency diagram shows central tendency.

EXAMPLE 14.5   For the same watershed in Example 14.4, maximum monthly flows were analyzed over the same 25 yr period using the method of Snyder and Wallace [4].

The data were experimented with until it showed that a class size of 50 cfs resulted in strong central tendency in the frequency diagram. The log normal probability density function was fitted to the frequency diagram using nonlinear least squares and the optimized parameters were mean = 2.39, and standard deviation = 0.380. The fitted density function is shown in Fig. 14-4.

Using Appendix I, the desired return periods for First Creek, Knoxville approximated by this method are compared with the previous method in in Fig. 14-5 and in Table 14-7.

Hence, the two methods are within 40% of one another.

Although there is no way to prove once and for all that one method is superior to another, there are distinct advantages of each method as compared in Table 14-8. The main reasons for the relatively close results of Examples 14.4 and 14.5 are the lack of outliers, to be discussed next, and the relatively long years (25 yr) of record, statistically speaking.

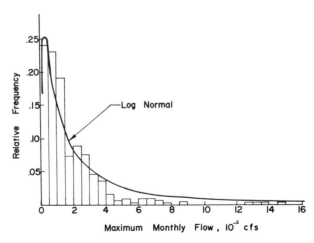

FIG. 14-4. Fitting maximum monthly flows for First Creek at Mineral Springs Av., Knoxville, Tennessee, using nonlinear least squares.

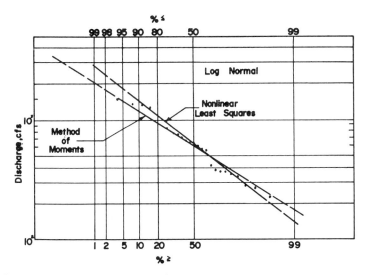

FIG. 14-5. Comparison of methods of moments with nonlinear least squares fitting for maximum annual flow on First Creek at Mineral Springs Av., Knoxville, Tennessee.

TABLE 14-7

	Design flow (cfs)	
Return period (yr)	Method of Example 14.4	Method of Example 14.5
100	2080	2800
50	1750	2350
33.3	1600	2100
25	1500	1920

TABLE 14-8 Comparison of methods of frequency analysis in Examples 14.4 and 14.5

Factor	Method recommended by Water Resources Council	Snyder–Wallace method (nonlinear least squares)
Data	Maximum annual	Can use maximum monthlys
Plotting position	Assumption required	Not required
Outliers	Censoring required	Censoring not required
Class size	Not required	Testing required
Computations	Manual	Computer required

14-6   Outliers

An outlier is a flow observation which is very much larger than the other flow observations. Including an outlier into the frequency analysis will result in a very poor fit of any cumulative distribution to the data. Not only will a poor fit result and severely bias the extrapolation, but a disproportionate amount of weight will be given to the outlier if objective fitting is done using the method of moments. Hence, it is customary to "censor" the data for outliers, meaning that certain statistical tests will be performed to determine if a very large observation relative to the remainder of the data is indeed an outlier. If it is, then the observation is removed from the data set or "censored."

If the density function is fitted to the frequency diagram made up of the observations, an outlier need not be censored because it is given its appropriate weight. Further, it is essential that the outlier be left in the data set for best definition of the character of the frequency diagram.

14-7   Confidence Intervals—Reliability

In order to properly qualify predictions made using frequency analysis, an index of the reliability of those predictions need be developed. A commonly used criterion is confidence intervals which provides quantified limits of the certainty associated with a prediction. For example, if the flood with a 100-yr return period is found by frequency analysis to be 2000 cfs, it normally is qualified by saying that we are 90% certain that this design flood is no more than $(2000 + \delta_1)$ cfs but no less than $(2000 + \delta_2)$ cfs.

The basis for establishing confidence limits is the *central limit theorem*, which states that for a population with finite variance σ^2 and mean μ, the sample means of repeated samples will be distributed as a normal distribution with mean μ and variance equal to σ^2/n where σ is the population standard deviation and n the population size. The theorem applies to all distributions; however, the distribution of sample means approaches a normal distribution as the sample size increases. The statistic $\sigma/\sqrt{n}$ is referred to as the standard error of the mean and it is apparent that reliability is a function of sample size.

Establishing reliability by directly utilizing the central limit theorem required a knowledge of the population statistics, whereas only the statistics of the sample are known. Thus, a different technique is needed for estimating confidence intervals for a sample mean and standard deviation. The technique utilizes the Student's t distribution which converges to the normal distribution as the sample size gets large. The approach then is to draw an inference from the sample as to the statistical characteristics of the population. The theoretical aspects of sampling techniques and hypothesis testing are beyond the scope of this text.

Beard [5] developed a method for establishing a reliability band above and below the fitted cumulative distribution. The factors by which the standard deviation of the random variate must be multiplied to establish a 90% reliability band are shown in Table 14.9. The 5% level means that 5% of future floods should be higher than this upper limit and 5% should fall under the 95% level. In other words, we are 90% certain that the 100 yr flood is within that band.

EXAMPLE 14.6   Establish a 90% confidence band on the fitted log normal distribution for the First Creek watershed in Example 14.4.

The 90% confidence band is plotted in Fig. 14-6. An example calculation is shown here. At the 1% exceedance frequency (100 yr discharge), the statistics are found from Table 14-9, and the 90% confidence band at this exceedance level is

$$5\% \text{ level} = \log(2080) + 0.825(0.237) = 3.513, \qquad Q_5 = 3260 \text{ cfs}$$

$$95\% \text{ level} = \log(2080) - 0.535(0.237) = 3.191, \qquad Q_{95} = 1550 \text{ cfs}$$

Thus, we are 90% certain that the 100 yr flow is no less than 1550 cfs and no larger than 3260 cfs.

TABLE 14-9   Error limits for flood frequency curves at 5 and 95% levels

Years of record (n)	Exceedance frequency (%), @ 5% level						
	90	10	50	1	0.1	99.9	99
5	0.76	2.12	0.95	3.41	4.41	1.22	1.00
10	0.57	1.07	0.58	1.65	2.11	0.94	0.76
15	0.48	0.79	0.46	1.19	1.52	0.80	0.65
20	0.42	0.64	0.39	0.97	1.23	0.71	0.58
30	0.35	0.50	0.31	0.74	0.93	0.60	0.49
40	0.31	0.42	0.27	0.61	0.77	0.53	0.43
50	0.28	0.36	0.24	0.54	0.67	0.49	0.39
70	0.24	0.30	0.20	0.44	0.55	0.42	0.34
100	0.21	0.25	0.17	0.36	0.45	0.37	0.29
	10	90	50	99	99.9	0.1	1
	Exceedance frequency (%), @ 95% level						

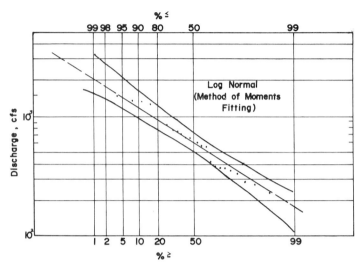

FIG. 14-6. 90% confidence band for maximum annual flow frequency distribution of First Creek, at Mineral Springs Av., Knoxville, Tennessee.

14-8 Effects of Urbanization on Stormwater Frequency

Espey and Winslow [6] reported a frequency analysis of 60 urban watersheds including 27 in Texas, 15 in Virginia, 8 in Maryland, 1 in Washington, D.C., 2 in Deleware, 4 in Mississippi, 2 in Michigan, and 1 in Illinois. The watersheds have experienced varied degrees of urbanization and associated channel improvements. The maximum length of record for any one watershed was 43 yr, and the average length of record was 14 yr and most of the study watersheds had a continuous record dating back from 1972.

Each maximum annual flood record was analyzed using the Log Pearson III distribution, the Weibull plotting position, and the method of moments fitting technique. From this analysis, flood frequency values for return periods of 2.33, 5, 10, 20, and 50 yr were determined. These frequency values were then correlated with four watershed and rainfall characteristics using stepwise multiple regression in the following model form

$$Q_i = aA^b I^c S^d R_i^e \Phi^f \tag{14-4}$$

where Q_i is a flood of return period i years, A the surface drainage area in sq. miles, I the percentage of impervious cover, S the main channel slope in ft/ft, R_i the 6-hr rainfall of return period i, and Φ a factor defined in previous studies which describes the roughness characteristics of the main channel,

Φ_1, and the extent of storm sewer improvement, Φ_2. The Φ factor is the sum of the two component factors as shown in Table 14-10.

The results of the stepwise regressions for all 60 urban watersheds is shown in Table 14-11. The coefficients of determination (using log transforms of the original variates) ranged from 0.92 to 0.98 and the average absolute error was 32%. The most important finding as demonstrated in Table 14-11 is that the channel urbanization factor becomes more significant and the percent impervious cover becomes less significant as the recurrence interval increases. This agrees with the popular view that at higher return periods, the rain intensity is so large that the pervious watershed cover acts essentially as an impervious cover.

Despite the relatively short periods of record for most of the urban watersheds in the study, the regression equations can be used as a basis for

TABLE 14-10   Classification of Φ

Values of Φ^a	Classification
For Φ_1	
1.0	Natural channel conditions
0.8	Some channel improvement and storm sewers; mainly cleaning and enlargement of existing channel
0.6	Extensive channel improvement and storm sewer system, close conduit channel system.
For Φ_2	
0.3	Heavy channel vegetation
0.2	Moderate channel vegetation
0.1	Light channel vegetation
0.0	No channel vegetation

$^a\Phi = \Phi_1 + \Phi_2$.

TABLE 14-11   Coefficients and exponents in flood frequency Eq. (14-4) for all 60 urban watersheds[a]

Return period (yr)	Exponents and coefficients					
	a	$b(A)$	$c(I)$	$d(S)$	$e(R_i)$	$f(\Phi)$
2.33	169	0.77	0.29	0.42	1.80	-1.17
5	172	0.80	0.27	0.43	1.73	-1.21
10	178	0.82	0.26	0.44	1.71	-1.32
20	243	0.84	0.24	0.48	1.62	-1.38
50	297	0.85	0.22	0.50	1.57	-1.61

[a] Studied by Epsey and Winslow [6].

drawing tentative conclusions regarding the flood frequency domain of small urban watersheds. However, application of the model of urban watersheds to the 50-yr return period should be done with caution and the predictions should be placed in the context of historical rainfall and streamflow data and tempered with good common sense.

The flood frequency predictive model of Espey and Winslow [6] was applied by those authors to two watersheds which were not part of the data base used in deriving their model. The average error associated with the predictions was 10 and 30%, respectively.

EXAMPLE 14-7   The model was used to predict flood frequencies for the First Creek Watershed at Mineral Springs Area, Knoxville, Tennessee. The following input data were determined for this watershed: $A = 12.5$ sq miles; $I = 0.20$; $\Phi = 1.0$; and $S = 0.01$. Using Hershfield's rainfall atlas, the following rainfall volumes were determined

$$R_{2.33} = 2.5 \text{ in.}, \qquad R_5 = 3.0 \text{ in.}$$

$$R_{10} = 3.5 \text{ in.}, \qquad R_{20} = 3.8 \text{ in.}, \qquad R_{50} = 4.5 \text{ in.}$$

Using Eq. (14-4) and Table 14-11, the following stormflows given in Table 14-12 were determined and compared with the results obtained from the frequency analyses in Examples 14-4 and 14-5. The results indicate that there is general agreement of the predictive model of the Epsey and Winslow model with the analyzed flood record. The results are much more sensitive to the very high and very low return periods. This is where the confidence band is largest and is associated with the necessity of extrapolating the record.

As more urban hydrologic data become available, the predictive equations can be tested and eventually the new data can be added to the data base for use in rederiving the model. Presently this model serves as a good indicator of the flood frequency values for ungauged urban watersheds.

TABLE 14-12

Return period (yr)	Design flow (cfs)		
	Example 14.4	Example 14.5	Espey and Winslow [6]
2.33	650	700	560
5	930	1090	780
10	1180	1420	1040
20	1410	1800	1320
50	1750	2350	1890

14-9 Design Risk

Once a design return period is selected, the next task is to select a design horizon over which to economically justify the project. This will involve calculating the probability that the design event will be equaled or exceeded over the design horizon.

For a design return period of T-years, the probability of that event being equaled or exceeded in any given year is

$$P = 1/T \tag{14-3}$$

and, the probability of the event not being equaled or exceeded in any given year is

$$q = 1 - (1/T) \tag{14-4}$$

Since the events are independent and mutually exclusive, the probability P_0 of the design event not being equaled in the first two years of the design period is

$$P_0(2) = q_1 \cdot q_2 \tag{14-5}$$

Therefore, the probability that the design event will be equaled or exceeded at least once in n years is

$$R = 1 - [1 - (1/T)]^n = 1 - P_0(n)^n \tag{14-7}$$

This is called design risk, and the longer the design horizon n, the higher the risk will be that the design event will be equaled or exceeded.

EXAMPLE 14-8 Calculate the chance that a discharge with a 25-yr return period will be equaled or exceeded (a) sometime in the 10-yr design horizon, (b) in any given year, and (c) exactly in the 5th year.

(a)
$$R = 1 - [1 - (1/25)]^{10} = 0.34$$

There is a 34% chance that the 25 yr flow will be equaled or exceeded sometime in 10 yr.

(b)
$$R = 1/25 = 0.04$$

There is a 4% chance that the 25 yr flow will be equaled or exceeded in any given year.

(c)
$$R = q_1 \cdot q_2 \cdot q_3 \cdot q_4 \cdot T^{-1} = 0.033$$

There is a 3.3% chance that the 25 yr flow will be equaled or exceeded exactly in the 5th year.

EXAMPLE 14-9   For the listed flood damages given as a function of return period, calculate the expected damages (a) associated with each of the return periods in each year of the design period (10 yr) and for the entire 10-yr design period, and (b) the total expected flood damages over the design period for the First Creek watershed in Knoxville, Tennessee. Assume a discount factor of 7% (see Table 14-13).

Expected flood damages over the next 10 yr are determined by

(a)   calculating the chance of each flood of rank i occurring exactly in each of the next 20 yr, j, $R(i, j)$,

(b)   multiplying the risk by the associated flood damage, $D(i)$, and dividing by the discount factor (1.07), and

(c)   summing the expected damages for each year and accumulating over the 20-yr design period.

The expected damages in any given year $FD(j)$, are

$$FD(j) = \sum_{i=1}^{10} R(i, j) \times D(i) \times (1.07)^{-j}$$

and, the total damages $TD(10)$ over 10 yr are

$$TD(10) = \sum_{j=1}^{10} \sum_{i=1}^{10} R(i, j) \times D(i) \times (1.07)^{-j}$$

For the problem at hand, the numbers were inserted into the equations and the result was $769,825 or an average annual expected damage of $76,982.

TABLE 14-13

Return period (yr)	Rank	Flood plain damages, 1972 ($)
100	1	2,500,000
50	2	1,220,000
33.3	3	890,000
25	4	532,000
20	5	261,000
16.7	6	105,000
14.3	7	70,400
12.5	8	20,900
11.1	9	5600
10	10	2100

14-10 Conclusions

The state of the art in stormwater frequency modeling is not as yet well developed. There is a strong need for uniformity in technique application, however, as yet there exists no scientifically based reasons either theoretical or empirical as to which set of techniques are best. However, there are brilliant minds hard at work on solving these problems and significant results will be forthcoming.

Although the frequency techniques presented herein were applied principally to floods, the same principles can be applied by analogy to a host of hydrologic phenomena including rainfall and water quality constituents. Water quality frequency under flood conditions could easily be estimated by establishing empirical relations between concentration of a conservative substance with flowrate (rating) and then determining the concentration of an associated return period of the flowrate.

Problems

14-1. What is the return period of a flood which has a 5% chance of being equaled or exceeded in any given year?

14-2. What are the principal statistical factors which affect frequency analysis of a flood record?

14-3. Distinguish between probability density function and cumulative distribution.

14-4. Prove that $f(x) = (1/a)e^{-x/a}$ is a probability density function. What is the mathematical form of the cumulative distribution?

14-5. Prove that the exponential function is a special case of the gamma function.

14-6. Prove that a and a^2 are the first two moments of the exponential function. Determine the first two moments of the uniform function.

14-7. What is the plotting position associated with an event of rank 3 in a 20 yr record assuming (a) the Weibull method, (b) the Beard Method?

14-8. There are 10 yr of continuous streamflow records available on Short Creek, Alabama (see Table 14-14). The drainage area is 22.7 sq.miles and the rainfall data are inadequate to apply a deterministic or parametric model. The land use has remained essentially constant during the 10 yr period of record. Using the Weibull plotting position, the log normal distribution and the method of moments,

(a) estimate the magnitude of the flood flows associated with a 100, 50, 5, and 2.33 yr return periods, and

TABLE 14-14

Year	Max. flow (cfs)	Year	Max. flow (cfs)
1975	4320	1970	990
1974	3590	1969	2120
1973	2560	1968	770
1972	1890	1967	1580
1971	2020	1966	860

(b)   establish the 90% confidence band around the distribution.

14-9.   There is extensive development planned for the Sixth Creek Watershed in Knoxville, Tennessee. Presently there is only 10% imperviousness and the stream channel ($S = 0.01$) is in natural conditions with heavy channel vegetation. After development, there will be 50% imperviousness, with an extensive channel improvement and storm sewer system. The channel will have light vegetation. Using the design rainfalls in Example 14-7, estimate percentage increase in maximum annual flows associated with return periods of 2.33, 5, 10, 20, and 50 yr associated with the proposed developments. As the city engineer, what measures would you propose to offset the increased flood flows in order to avoid lawsuits by downstream riparian owners?

14-10.   Estimate the chance that a flood flow with a 100-yr return period will be equaled or exceeded (a) in any given year, (b) exactly in the next 25 yr, and (c) sometimes in the next 100 yr.

14-11.   Rework Example 14-9 assuming no discount factor and compare the results.

References

1. Viessman, W., Jr., Harbaugh, T. E., and Knapp, J. W., "Introduction to Hydrology." Intext, New York, 1972.
2. U.S. Water Resources Council, A uniform technique for determining flood flow frequencies, Bull. No. 15 (1967).
3. Chow, V. T., Statistical and probability analysis of hydrologic data. In "Handbook of Applied Hydrology" (V. T. Chow, ed.). McGraw-Hill, New York, 1964.
4. Snyder, W. M., and Wallace, J. R., Fitting a three-parameter log-normal distribution by least squares, Nordic Hydrol. **5,** 129–145 (1974).
5. Beard, L. R., Statistical methods in hydrology, Civil Works Investigations, Sacramento District, US Army Corps of Engineers, 1962.
6. Espey, W. H., Jr., and Winslow, D. E., Urban flood frequency characteristics, J. Hydr. Div., Amer. Soc. Civil Eng. **100** (HY2), 279–293 (1974).

7. Espey, W. H., Jr., Morgan, C. W., and Masch, F. D., A study of some effects of urbanization on storm runoff from a small watershed, Center for Research in Water Resources, Univ. of Texas, Austin, Texas, July 1965.

8. Espey, W. H., Jr., and Winslow, D. E., "The effects of urbanization in unit hydrographs for small watersheds, Houston, Texas, 1964–1967," TRACOR Document Nos. 68-975-U and 68-1006-V, Office of Water Resources Research, Oct. 1968.

9. Hershfield, D. M., Rainfall frequency atlas of the United States, US Weather Bureau Tech. Paper No. 40, 1961.

PART V | Stormwater Quality Modeling

Chapter 15 | State of the Art in Stormwater Quality

15-1 Introduction

To insure an adequate supply of high quality water, management of the global water resources is required. These management policies should allocate upstream waste loadings such that the waste assimilative capacity of the receiving water body is not overloaded, and the specified quality for downstream consumptive use is met. Until recently, essentially all plans for providing downstream water quality have assessed the impact and subsequently set controls only on discharges from known sources, the so-called point sources. It has been popularly assumed that the pollution contributions of stormwater represented only a background, or natural pollution source and was of minor pollution potential. But now, in light of the results from the few, but very good, recent studies qualifying and quantifying the pollution potential of stormwater, it is obvious that the impacts of both the point and nonpoint pollution sources on downstream water quality must be assessed if adequate management policies are to be formulated (see Fig. 15-1).

Stormwater quality varies widely but is not unpredictable. The nature and extent of pollutants present at a given runoff rate vary both from storm to storm and with the season of the year, as well as within an individual

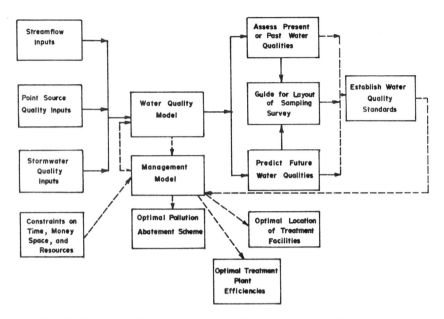

FIG. 15-1. Schematic diagram for water quality management studies, showing inputs and outputs.

storm event. The land use pattern on which stormwater is generated basically governs the type and concentration of the pollutants present. As discussed in earlier chapters land use also influences the quality of stormwater.

The principal land uses discussed in this chapter are: urban, agriculture, mining, and forests and woodlands. The discussion is broken down further according to human activities within each land use area. This chapter is a prelude to the following chapters on stormwater quality modeling. Available data coupled with modeling provide the inputs for assessment studies.

15-2 Stormwater as a Pollutant

A pollutant is "something" added to the water resources that impairs natural water quality making the water unsuitable for potential downstream consumptive use. Typical pollutants are human and animal excreta, bacteria, sediment from soil erosion, oils, metals, nutrients, and toxicity. All of these pollutants are generated by humans and their activities, and are found in concentrated waste streams and dispersed in varying quantities on the land. Stormwater serves as a vector for these and other pollutants.

Rainfall and the associated runoff loosen, suspend, and then transport those pollutants over the land surface to receiving water bodies. Has everyone noticed how "dirty" the creeks and rivers become during runoff from storm events? Stormwater, per se, is not a pollutant, but the presence of the many suspended pollutants gives stormwater a significant pollution potential.

The "first-flush" effect of many runoff events may constitute a shock load to receiving streams, particularly during the low flow summer months. The term "first-flush" applies to the beginning of a runoff event when the volume of overland flow is still small, but the concentration of pollutants is very high. Those pollutant particles lying loose on impervious surfaces are quickly suspended and transported by the incipient overland flow. Pollutant concentrations are observed to decrease after the "first-flush" of the loose particles until higher volume flows scour compacted particles.

For too long it has been accepted that rainfall and the consequent increased flow in streams somehow "diluted" pollution. Recent studies, however, have shown that pollution is no less, and often even worse, during these periods of increased streamflow. Runoff from agricultural lands carry large amounts of sediment as well as nutrients. Studies of urban stormwater quality have established that many events are, at least, as strong in organic pollution (BOD) as the discharge from a sewage treatment plant with secondary treatment [1, 2]. The same flows were found to have a substantially higher total solids content than would be expected from average *raw* domestic wastewater. Hence, stormwater is allegedly a significant source of stream and lake pollution.

15-3 Nonpoint Source

Pollution sources are classified either as point or nonpoint. Point sources are waste discharges at discrete points. The origin of the waste stream is either known or readily identified. Monitoring and control of a point source discharge is accomplished in-line of the waste stream, be it a pipe or open channel. Examples of point sources of pollution are the discharges from sewage treatment plants.

Nonpoint source pollution is generated from the various land use areas. The different human activities within these areas generate pollutants that accumulate on the land, the streets, etc. Rainfall and the associated runoff remove many of these pollutants and convey them ultimately to the receiving streams and lakes. Thus, nonpoint source pollution is simply stormwater quality.

Nonpoint source pollution is described as:

1. Diffuse, not concentrated at discrete points;
2. Intermittent, not flowing continuously;
3. Arising from extensive areas, and in transit over land;
4. Responsive to climatic conditions.

Also, flows from the nonpoint sources cannot be monitored at any point of origin because exact sources are not always clearly identified. Any prevention or control is site specific through land use changes and controls. Treatment is possible, though possibly expensive, after flows from contributing areas have been collected in discrete waste streams.

15-4   Effects of Pollutants on Receiving Streams and Lakes

To assess properly the impact of storm runoff on the water quality of receiving streams and lakes, the student, the engineer, the researcher, and the planner must have an appreciation and understanding of the impacts of the individual classes of pollutants. The various pollutants affect differently the qualities, the biota, and the consumptive uses of the receiving water bodies. However, one must remember that *pollution of the water resources is not just the cumulative effects of the individual pollutants.* The presence of two or more of the different pollutants in the water may intensify or decrease their cumulative pollutional effect. One example is the effect of heavy metals on the decomposition of organic matter. Aerobic bacteria utilize the dissolved oxygen in the breakdown of the organics causing an oxygen deficit. The impact of an oxygen deficit might be a fish kill. Heavy metals are a form of toxicity to the bacteria and act to inhibit their decomposition activities.

Pollution of the global fresh water resources almost always affects adversely any potential downstream consumptive uses. Polluted water

1. poses a threat to public health, particularly in the case of consumption or body contact;
2. increases the costs of treatment for downstream water supplies;
3. may not be suitable as industrial process water;
4. may not be suitable as agricultural irrigation water;
5. impairs or destroys the aquatic environment and biota; and
6. is not acceptable aesthetically.

The following classes of pollutants are viewed as being those pollutants most likely to be in stormwater. A brief discussion of the most common sources and effects on receiving water bodies is offered for each pollutant class.

Organic Wastes The principal sources for organic wastes in storm-water are human and animal excreta, decaying plant and animal matter, and indiscriminately discarded litter and food wastes. The principal elements composing organic compounds are carbon, hydrogen, oxygen, and nitrogen. In the presence of sufficient oxygen, aerobic bacteria assimilate the organic substrate as their energy source, reducing the organics to carbon dioxide, water, and ammonia. This stabilization process utilizes the stream- or lake-dissolved oxygen resulting in an oxygen deficit. (A deficit is any oxygen content in the water less than the saturation concentration.) The dissolved oxygen content of a stream or lake is one of the prime indicators of the health of that body of water. If the magnitude of an oxygen deficit is great enough a fish kill may occur. A secondary demand on the dissolved oxygen is the oxidation of the ammonia to nitrites and nitrates, i.e., the nitrogenous oxygen demand.

In addition to being a sink for the water's dissolved oxygen, the assimilation of the organics provides the energy source for high bacterial populations. This creates a water unfit for human contact or consumption.

Bacteria The extent of bacterial contamination is perhaps the single most important indicator of water quality, especially water intended for human consumption or body contact. The principal sources for bacteria are human and animal excreta, decaying plant and animal matter, and soil. For almost a century, such indicators as the coliform bacteria have been used to monitor fecal contamination of water resources. As discussed above, the decomposition of organics provides the energy source to support the growth of high bacterial populations. The concern over bacteria relates to the transmission of disease.

Sediment Typically defined as eroded soil, sediment includes earth material ranging in size from clay in colloidal suspension to boulders. The nature and volume of sediment is a function of land use practices. Such human activities as construction, row-cropping, surface mining, and forest clear-cutting disturb the soil and its protective cover. The exposed soil material is easily suspended and transported by storm runoff.

Sediment is a stormwater pollutant of major importance. Stormwater appears muddy due to suspended clay particles. As the velocity of flood water decreases, the sediment falls out of suspension, blanketing the channel bottom. Benthal communities may be covered and destroyed. A high content of suspended solids in water is injurious to the gills of fish. The colloidal soil particles create a problem with color and turbidity and may increase treatment costs for downstream water supplies. Also, sediment is a major vector for pesticides and inorganics, particularly phosphorus. Pesticides are "present with the soil" due to the nature of their application.

Phosphorus, a nutrient for the eutrophication process, is bound chemically with the soil particles.

Nutrients   Nitrogen and phosphorus are the principal nutrients carried by stormwater. The principal sources of nitrogen are organic wastes and fertilizers, and the principal sources of phosphorus are sediments and fertilizers. Nutrients in receiving water bodies are the primary cause of eutrophication.

Metals   Heavy metals are conservative substances that may undergo biological magnification in the food chain. At increasing concentrations, many of the metals become toxic, either to man, livestock, and wildlife, or to any of the many aquatic organisms. Metals inhibit bacterial activities in decomposing organic wastes. As is true with most pollutants, the principal sources are created through human activities and land use patterns. The significant sources of metal contamination of stormwater are streets and roadways, and industrial sites.

Oil and Other Petroleum Products   Oils and other petroleum products are used for fuel and lubricants at all levels of human activity. The by-products from the combustion of gasoline and oil accumulate on streets and buildings as dustfall. Oil and grease are spilled from machinery and automobiles. These pollutants are readily transported by storm runoff to receiving streams.

Oil floats on water and forms a film that impedes the natural re-aeration process, i.e., the diffusion of oxygen from the atmosphere into the water. Any oils in water interfere with the life processes of aquatic biota, including waterfowl. Also, oils contribute a chemical oxygen demand (COD). COD is a measure of the total oxygen required to convert organics to water and carbon dioxide. COD exceeds BOD in that some of the organics can not be decomposed solely by bacterial action (BOD).

Pesticides   Pesticides are used in conjunction with many land use activities. Farmers apply insecticides and herbicides to crops, electric companies use herbicides to keep down tree growth under power transmission lines, home gardeners use insecticides to save their vegetables. By far, the major use is in agriculture. Pesticides are very persistent in nature. Some varieties take many years to decompose and, like heavy metals, may undergo biological magnification in the food chain. Decomposition products may prove to be even more toxic or carcinogenic than the parent compounds. After application, the pesticides lie exposed on the area of application and are subject to the erosive forces of rainfall and runoff. The concern for pesticides relates to their toxic and carcinogenic effects.

pH and Alkalinity   pH is a term used to express intensity of the acid or alkaline condition of a solution [3]. The alkalinity of water is a measure of its capacity to neutralize acids. Acid and alkaline waters reflect their source and certain chemical compounds present in solution. One example is the acid mine drainage resulting from the oxidation of pyritic materials. In the strip coal mining region of Tennessee, the acid drainage is soon buffered by carbonate from the abundantly present limestone. Another example of acidic stormwater is the runoff of rain-made acid by combination with SO_x present in the atmosphere from the burning of high sulfur coal at thermal power plants. The pH and alkalinity of water affects both the survival of aquatic biota and human use of that water.

15-5   Quality of Urban Runoff

In the United States less than 3% of the land area is classified as urban, but more than 70% of the population is concentrated in this small segment [4]. On a daily basis, the many activities associated with such high density population centers require extensive amounts of natural resources, fuels, land resources, and foodstuffs resulting in the production of large volumes of waste materials. The wastes are either contained in discrete waste streams and collection stations, discharged to the atmosphere, or dispersed on the land surface. Those waste items that are spread over the land, plus dustfall from the atmosphere, generate the water-borne pollutants in urban stormwater. Due to the concentration of population, commerce, and industry, and the consequent production and accumulation of significant amounts of nonpoint source pollution, the quality and impact of urban runoff are of major importance to the proper management of water.

A typical urban setting has areas for commerce and industry surrounded by residential areas, and a complex of streets and roadways connecting these interdependent parts. The numerous rooftops and paved surfaces are impervious areas that, in many cases, comprise more than one-half of the total urban area. Currently, land use patterns within many urban areas are changing daily. Pervious areas are being converted to impervious areas through the construction of new roads, residences, and commercial and industrial buildings. Impervious zones generate larger volumes of runoff and have associated shorter times to equilibrium for a given rainfall. In fact, *rainfall that generates measurable runoff from an impervious zone may not generate any runoff from a pervious zone*. The impacts of the high degree of imperviousness in urban areas are the more frequent occurrence of storm runoff and a greater percentage of the accumulated pollutants washed to the receiving waters.

Five principal land use patterns are identifiable in all urban settings. They are construction, industry, commerce, streets and roads, and residential sections. The following paragraphs give brief discussions of the individual land uses, identify the prominent human activities associated with each and the characteristic pollutants generated by those activities. The reader must bear in mind that these are not offered as detailed descriptions, but only as generalities designed to help the investigator identify sources and characterize pollutants in his study of urban stormwater runoff. It is hoped that the reader will develop an appreciation of the varying effects of different land uses on stormwater quality. It is important that planners and engineers realize these effects when they assess the impacts of proposed land use changes.

Construction   Construction involves, first, site preparation, and second, the "construction" of the desired end product, whether it is a road, sewer, or building. Site preparation includes "clearing and grubbing" of a new site, dynamite blasting of rock, demolition of existing structures at an old site, excavation, "cut and fill" of earth to proper grade, compaction of foundation, and the hauling of excess materials. During the construction phase, proper building materials and machinery, in addition to a labor force, are employed to "build" the specified end product.

The pollutants generated during construction are connected directly to the on-going construction activity. During the early site preparation the protective cover is removed and the soil disturbed. Rainfall and consequent runoff erode the loosened earth material and the sediment from this soil erosion is the major pollutant identified with construction. In addition to suspended solids, color, and turbidity, sediments also are a source of nutrients and heavy metals that were bound with the soil. The disturbance of the soil generates dust that becomes entrained in air currents and is dispersed over the construction site and the neighboring areas. Some of the pollutants generated during the construction phase are oils and lubricants spilled from machinery, metals and inorganics leached or corroded from building materials, and organics originating with litter and food wastes.

Industry   Many industrial activities contribute significantly to stormwater pollution. Rainfall and runoff leach pollutants from open stockpiles of raw materials, finished products, and process wastes. The spills during the handling of materials and the leakage from piping systems and corroded storage units add, in part, to the accumulation of pollutants on exposed surfaces. Process gases contribute stormwater pollutants either through dustfall or washout. The nature of pollutants from an industrial site is a function of the type of industry. The pollutants that could have serious impact on receiving water bodies include oils, toxicity, and heavy metals.

Commerce Most commercial areas are highly impervious, being comprised of buildings surrounded by parking lots and streets. The pollutants that accumulate on the exposed surfaces are generated by the extensive automobile traffic, litter, dustfall, and spills. The pollutants associated with commercial areas are essentially the same as the pollutants associated with streets and roads.

Streets and Roads Following their construction, urban streets and roads are used exclusively for car, bus, and truck traffic. Pollutants accumulate from dustfall, spills, littering, and wastes generated during automobile use. The pollutant sources associated with automobile use include exhaust emissions, tire wear, and leakage of oil and lubricants. These sources generate, respectively, such pollutants as lead, rubber compounds, oils, and COD. Most streets and roads have been paved with asphalt or concrete, creating a highly impervious surface. A great percentage of the pollutants that fall onto the streets are swept into the gutters by wind from the wake of passing automobiles. Any runoff from the street surface suspends these loose particles and quickly transports them to receiving streams. For this reason the "first-flush" effect is observed most often in runoff from streets.

Residences The quantities of pollutants generated in residential areas vary with the population density, the amount of open space, and the general standard of living. The population density affects waste production because more people generate more waste; however, this axiom does not always hold. Because of the different standards of living, more often those living in the less densely populated areas can afford more goods and services, and therefore, have greater opportunity for waste production. The standard of living also reflects the general cleanliness of a neighborhood. In this respect, those people in impoverished areas seem to allow greater accumulation of pollutants on their yards and streets. Generally, more open space is significant in reducing the volume of runoff and the amount of pollution transported to receiving streams.

Organics, bacteria, and nutrients are the most common pollutants in stormwater from residential areas. Their sources are litter, tree leaves, and human and animal excreta. During the early years of a new residential division, unfinished construction will contribute sediments. Also, measurable amounts of pollutants are generated by dustfall and automobiles.

It is obvious from this discussion of the principal sources of pollutants within the urban land use categories that dustfall may contribute significantly to the pollution of urban stormwater. Studies have demonstrated that a definite correlation exists between dustfall and urban stormwater quality [5, 6]. The pollutants in dustfall may be either soluble or insoluble in water.

A list of the pollutants includes dirt, nitrates, calcium, sulfates, phosphates, lead, and organics. These pollutants originate from industry, fuel combustion, automobile emissions, pollen, and even small dirt and dust particles that were entrained in air currents. The air currents mix the various pollutants and disperse them over wide spatial areas. They are removed from the air either by natural fallout or by washout during precipitation.

The characteristics of urban stormwater quality are given in Table 15-1. The results of studies from three different urban areas and a general range of values reported by the American Public Works Association are presented. Note the wide range of concentrations for each pollutant; the concentrations are known to vary with the seasons of the year, on-going human activities, time since the last runoff, and duration and intensity of rainfall. Due to the probabilistic nature of rainfall and the difficulty and expense associated with extensive sampling programs, most study groups assess only average concentrations of the storms.

Intuitively, different land use practices will generate different types and amounts of pollutants. As pointed out earlier, it is essential that the impacts of land use changes on the qualities of both stormwater and receiving streams be assessed before any change is made. Any stormwater management scheme selected "after the fact" necessarily will be from a narrower range of choice and quite probably will cost considerably more money. Properly planned management policies will insure optimum use of the water for the local users as well as those users downstream. Tables 15-2 and 15-3 show the differences in the rate of pollutant buildup with respect to different urban land use. These tables also show the differences in buildup with respect to different geographic locales.

15-6 Quality of Agricultural Runoff

On the basis of land area, agriculture stands as a potentially large contributor of pollution. The land area in the 50 states of this country is 2264 million acres [11]. Of this total, 1064 million acres, 47% are classified as "land in farms," which include cropland, grassland and pasture, and farm woodlots and roads. Cropland and pasture occupy 36 and 50.7% of the farmland, respectively. The major pollutants generated by agricultural activities are sediments, organics, nutrients, and pesticides. Agricultural land use includes cropland, grassland and pasture, and animal feed lots. A brief discussion, that identifies farming practices and associated pollutants, is given for each land use category.

Cropland Croplands are the largest producer of sediment [12]. Some cropland is tilled following fall harvest and lies exposed during the wet

TABLE 15-1 Characteristics of urban stormwater

Pollutant	Durham, North Carolina[a] Range (mg/l) Low	High	Mean	Cincinnati, Ohio[b] Range (mg/l) Low	High	Mean	Tulsa, Oklahoma[c] Range (mg/l) Low	High	Mean	APWA[d] Range (mg/l) Low	High
BOD	—	—	—	2	84	19	8	18	11.8	1	700
COD	20	1042	170	20	610	99	42	138	85.5	5	3100
Total solids	194	8620	1440	—	—	—	199	2242	545	450	14,600
Volatile suspended solids	5	970	122	1	290	53	—	—	—	12	1600
Suspended solids	27	7340	1223	5	1200	210	84	2052	367	2	11,300
pH	—	—	—	5.3	8.7	7.5	6.8	8.4	7.4	—	—
Total PO_4	0.2	16	0.82	0.07	4.3	0.8	0.54	3.49	1.15	0.1	125
NO_3	—	—	—	0.1	1.5	0.4	—	—	—	—	—
Chlorides	—	—	—	—	—	—	2	46	11.5	2	25,000
Pb	0.1	2.86	0.46	—	—	—	—	—	—	0	1.9
Ca	1.1	31	4.8	—	—	—	—	—	—	—	—
Cu	0.04	0.50	0.15	—	—	—	—	—	—	—	—
Fe	1.3	58.7	12	—	—	—	—	—	—	—	—
Mg	3.6	24	10	—	—	—	—	—	—	—	—
Mn	0.12	3.2	0.67	—	—	—	—	—	—	—	—
Alk	24	124	56	10	210	59	—	—	—	—	—
Fecal coliforms (#/100 ml)	100	200,000	23,000	—	—	—	10	18,000	420	55	112×10^6

[a] After Colston [1].
[b] After Weibel et al. [7].
[c] After AVCO Economic Systems Corporation [8].
[d] After American Public Works Association [9].

TABLE 15-2   Average daily loads per mile of street (Tulsa, Oklahoma)[a]

	Average load: lb/day/mile of street				
Land use	BOD	COD	Total solids	Organic Kjeldahl nitrogen	Soluble orthophosphate
Residential	1.98	13.9	63.1	0.14	0.18
Commercial	3.06	20.3	95.7	0.23	0.24
Industrial	3.51	27.7	354.	0.26	0.57

[a] After AVCO Economic Systems Corporation [8].

TABLE 15-3   Average daily loads per mile of street (Chicago, Illinois)[a]

Land use	Average load: lb/day/mile of street			
	BOD	COD	N	PO4
Single family residential	0.36	2.95	0.03	0.004
Multiple family residential	0.87	9.70	0.15	0.012
Commercial	2.70	13.6	0.14	0.024
Industrial	1.45			

[a] After Roesner [10].

winter months. That land not tilled in the fall is plowed early the next spring in time for planting season. In those areas of wheat, oat, and barley production, a winter cover crop is planted in the fall for a spring harvest. Whatever the crop production, the land is heavily disturbed at least once each year. This pulverizes the soil into fine, loose particles ideal for crops; but, it leaves the soil easily erodable by storm runoff. After the crop begins to grow, cultivation, which keeps the soil loose, may be required for weed and grass control. When the erosion occurs, the loose dirt is suspended and washed to receiving streams by the stormwater. Carried with the sediment are fertilizers, soil bacteria, plant matter, and pesticides. The erosion of sediment not only affects water quality, it represents the loss of a valuable natural resource—fertile soil for food production.

Both fertilizer and pesticides currently are used heavily in conjunction with crop production. The fertilizer is applied to the soil to provide the necessary soil nutrients required for crop growth. Most fertilizers contain the three basic elements: nitrogen, phosphorus, and potassium. Additionally, nitrogen is added to the soil through the fixation of atmospheric nitrogen by the legumes. The pesticides most used are herbicides and insecticides. Herbicides are used for the control of noxious weeds and unwanted grass, and pesticides are used for insect control.

After the crop has been harvested, residual plant matter remains. This may be corn stalks or wheat stubble; but whatever, it constitutes a potential for water pollution. The plant residue contributes BOD and nutrients, in addition to being a source of visible floating matter.

The results of studies [13] that measured the concentrations of pollutants in stormwater from a 173-acre cultivated site in eastern South Dakota are given in Table 15-4. The BOD values are small relative to the COD indicating the contributions of plant matter that was not easily decomposed by bacteria, as would be true of cellulose materials. The high suspended solids and nutrient concentrations suggest events with a high degree of erosion. Interpretation of the data is clouded without information on the intensity and duration of the individual rainfall events, and the volume of runoff.

Grassland and Pasture Grasslands and pastures provide grazing for animals. A wide choice of grasses, clovers, and legumes may be grown. Because the land is seldom disturbed by cultivation and a permanent cover crop is maintained, runoff from these areas is a minimum. Animal wastes and plant matter are the principal pollutant sources. Those pollutants include organics (BOD) and, to some extent, nutrients. Very little sediment should ever originate from grasslands and pastures due to the protection provided by the cover crop. This is demonstrated in Table 15-5. The data are also from a South Dakota study that monitored the quality of runoff from different agricultural land use areas [13].

Feedlots Animal feedlots are small areas where large numbers of animals are penned for the purposes of controlled intensive feeding. Some people may consider feedlots point sources rather than non-point sources of pollution. If all flows, both waste and storm, are *controlled* and discharged

TABLE 15-4 General quality of surface runoff from cultivated agricultural lands—results of studies on a 173-acre watershed in South Dakota[a]

Pollutant (mg/l)	Study 1	Study 2
BOD	5–30	3–15
COD	50–360	70–780
Suspended solids	90–5000	180–6000
Total phosphorus	0.26–2.4	0.04–0.60
Organic N + NH_3	1.3–20.3	2.8–17

[a]After Dornbush *et al.* [13].

TABLE 15-5 Average concentration of pollutants in agricultural runoff by land use[a]

Pollutant (mg/l)	Cultivated land	Pasture	Alfalfa and brome grass
Total residue	1241	222	108
Suspended solids	1021	38	40
Total phosphorus	1.05	0.49	0.35
Nitrate (mg/lN)	1.5	0.4	0.3
Total Kjeldahl nitrogen (mg/lN)	2.6	1.7	0.8
COD	148	49	22

[a] After Dornbush et al. [13].

at *discrete* points, feedlots are *point sources*. However, they must be considered *nonpoint sources* if any flows are *uncontrolled* and discharge from the area via "natural" runoff routes.

Animal excreta is a source for BOD, suspended solids, dissolved solids, and nitrogen compounds. The organic pollution potential of various farm animals relative to man is given in Table 15-6. Animal urine is the principal source for the ammonia that abounds at feedlot operations. Accumulated wastes that have sufficient moisture present form organic acids during anaerobic decomposition. Perhaps the most significant pollutant generated at feedlots is bacteria. The vast amount of organic waste provides the energy source that supports a high rate of bacteria growth. In a study of runoff from cattle feedlots, Miner *et al.* [15], reported bacterial densities constantly in excess of one million organisms per 100 ml.

15-7 Quality of Runoff from Mining Areas

To date, mining operations have affected only about 0.5% of the land area in the United States [12]. Even though this represents a very small

TABLE 15-6 Summary of adopted population-equivalents-carbonaceous BOD[a]

Species	Population equivalent	
	Human excrement baseline	Domestic sewage baseline
Man	1.0	0.55
Cattle	6.4	3.5
Hen	0.32	0.17
Sheep	0.57	0.31
Swine	1.6	0.9

[a] After Henderson [14].

amount of disturbed land relative to the acreage affected by agriculture and urban use, the impacts of mining are great. There are fundamentally two types of mining: surface mining and deep mining. All aspects of surface mining have direct impact on stormwater qualities; whereas, the only deep mining activities that contribute pollutants are initial excavations, hauling, and waste (gob) piles. Various mining operations include mining for coal, copper, bauxite, uranium, etc., and sand and gravel operations.

Surface Coal Mining In those areas where coal seams lie in close proximity to the land surface, surface mining techniques are employed. Large earth moving equipment removes the dirt overburden to gain easy access to the coal. The coal is then loaded onto trucks and hauled from the mining areas over dirt and gravel haul roads. These activities remove protective vegetable covering and disturb the soil, creating a source of potentially large sediment yield. In the mountainous Appalachian region of the United States, traditional contour surface mining practices have involved the dumping of overburden on the downslope side of the cut while exposing the coal seam. The streams draining these areas are literally filled with sediment.

Coal mining generates acid mine drainage, iron, sulfates, and certain mineral pollutants. Acid mine drainage originates from the oxidation of sulfide materials, such as iron sulfide. The mining operations uncover these sulfide compounds that are present with the coal and exposes them to atmospheric oxygen. Upon exposure to oxygen in a moist environment, the iron sulfide is converted in a three-step process to ferric hydroxide and sulfuric acid. Ferric hydroxide is the red precipitate that one sees blanketing the channel bottoms of streams that drain coal mining areas. Table 15-7 gives criteria for determining acid mine drainage. The impact of acid mine drainage on the aquatic biota is disastrous.

Other Mining Operations The most significant stormwater pollutant generated by the other mining operations, including sand and gravel oper-

TABLE 15-7 Criteria for determining acid mine drainage[a]

pH	less than 6.0
Acidity	greater than 3 mg/l
Alkalinity	normally 0
Alkalinity/acidity	less than 1.0
Fe	greater than 0.5 mg/l
SO_4	greater than 250 mg/l
Total suspended solids	greater than 250 mg/l
Total dissolved solids	greater than 500 mg/l
Total hardness	greater than 250 mg/l

[a] After Herrick and Cairns [16].

ations, is sediment. All mining activities that involve excavation create the potential for sediment production. In certain mineral mining industries, the processing of raw materials to concentrate ore creates vast piles of pulverized raw materials, called tailings, that are sources for sediment. In the western United States tailings piles resulting from hard rock mineral mining are spread over wide expanses of land. Additional pollutants such as oils may result from the operation of the mining machinery.

15-8   Quality of Runoff from Forests and Woodlands

Direct surface runoff is a minimum from forests and woodlands. The trees and their leaves intercept a large portion of the rainfall negating the impact of raindrops on the soil. What rainfall is unhindered by trees quickly infiltrates through the humus layer into the soil. Only during the most intensive rainfall is any surface runoff generated. The only significant pollution generated by established forests and woodlands are organics (BOD and COD) contributed by leaves, nutrients, and visible floating matter. In some areas, insecticides may appear in runoff following their use to control invading insects.

Logging operations, especially clear-cutting, remove the protective cover and, to some extent, disturb the soil. This creates the potential for sediment production. Haul roads are also sources for sediment, particularly those that travel directly downslope on a forested hillside. Associated with the sediments are organics, nutrients, and soil bacteria.

15-9   Conclusion

Stormwater is a potentially serious pollution threat to water in the environment. Rainfall and runoff "cleanse" the environment of the uncontrolled waste from human activities that is dispersed over the land, buildings, and streets. Due to the probabilistic nature of rainfall, stormwater is highly variable both in time and space. The impact of this intermittent pollution input to the receiving water bodies must be assessed if the quality of the water resources is to be maintained and improved. The few studies of stormwater quality that have been conducted to date have concluded that storm runoff is often detrimental to the receiving streams. Particularly during the dry summer months, stormwater is a shock load to the receiving water bodies and not always a dilution for pollution.

It is meaningful when the quality of stormwater can be related to land use practices and storm characteristics. Abatement and control of pollution due to storm runoff must consider not only concentration of the flows

for treatment, but also, land use changes and control. Presently, most of the available data are for quality variables only and do not include information about the associated storm event. Desirable storm information includes intensity of rainfall, duration of rainfall, volume of runoff, and time since the last runoff event. The capability to predict storm flow and quality for selected storms would allow the scientific assessment of the impact on receiving streams and an evaluation of the trade offs between different management policies.

Problems

15-1. Explain the meaning of the expression "stormwater is a pollutant."

15-2. Name the consumptive demands that you personally place on a local stream or lake, and explain how a strong organic waste loading would affect your use of this water body.

15-3. You represent the planning agency for a town having a population of 100,000 people. A new sewage treatment plant with tertiary treatment is proposed that will replace the town's two old and overloaded treatment plants. This new plant will discharge the treated wastewater into the local river that flows through the town. A political interest wishes the plant to be located downstream of the town. The local regulatory agency and EPA have established a set of stream water quality criteria that should not be violated. What role should the stormwater quality from the town play in your decision on where to locate the new plant? Do you think the impact of the urban runoff on the receiving stream quality should be assessed prior to the final decision? Explain.

15-4. Discuss how stormwater quality might change if a previously wooded area is cleared and a shopping center is built.

15-5. In your opinion, what separates "natural" water quality and non-point source pollution?

15-6. Explain how controlled land use changes might improve the quality of storm runoff.

References

1. Colston, N. V., Jr., Characterization and treatment of urban land runoff, *Environ. Protection Technol. Series,* EPA-670/2-74-096, December 1974.
2. Tafuri, A. N., Pollution from urban land runoff, Advanced Waste Treatment Research, U.S. EPA, April 11, 1975.
3. Sawyer, C. N., and McCarty, P. L., "Chemistry for Sanitary Engineers." McGraw-Hill, New York, 1967.

4. Ehrlich, P. R., and Ehrlich, A. E., "Population, Resources, Environment." Freeman, San Francisco, 1972.

5. Johnson, R. E., Rossano, A. T., Jr., and Sylvester, R. O., Dustfall as a source of water quality impairment, *J. Sanitary Eng. Div., ASCE* **92,** No. SA1, 245–267 (February 1966).

6. Barkdoll, M. P., An analysis of urban stormwater quality and the effects of dustfall on urban stormwater quality: East Fork of Third Creek, Knoxville, Tennessee. M.S. Thesis, Environ. Eng. Program, Dept. of Civil Eng., University of Tennessee, 1975.

7. Weibel, S. R., Anderson, R. J., and Woodward, R. L., Urban land runoff as a factor in stream pollution, *J. WPCF* **36,** No. 7, 914–924 (July 1964).

8. AVCO Economic Systems Corporation, Stormwater pollution from urban land activity, Water Pollution Control Research Series, Federal Water Quality Administration, Report No. 11034 FKL, July 1970.

9. American Public Works Association, Water pollution aspects of urban runoff, Water Pollution Control Research Series, Federal Water Quality Administration, Report No. WP-20-15, January 1969.

10. Roesner, L. A., Quality aspects of urban runoff. *In* "Short Course on Applications of Storm Water Management Models," University of Massachusetts Dept. of Civil Engr., August 19–23, 1974.

11. US Department of Agriculture, "Agricultural Statistics—1974."

12. US Environmental Protection Agency, Office of Air and Water Programs, Methods for identifying and evaluating the nature and extent of nonpoint sources of pollution, Report No. 430/9-73-014, October 1973.

13. Dornbush, J. N., Anderson, J. R., and Harms, L. L., Quantification of pollutants in agricultural runoff, Office of Research and Development, US Environmental Protection Agency, Report No. 660/2-74-005, February 1974.

14. Henderson, J. M., Agricultural land drainage and stream pollution, *J. Sanitary Eng. Div. ASCE* **88,** No. SA6, 61–74 (November 1962).

15. Miner, J. R., Lipper, R. I., Fina, L. R., and Funk, J. W., Cattle feedlot runoff—its nature and variation, *J. WPCF* **38,** No. 10, 1582–1591 (October 1966).

16. Herricks, E. E., and Cairns, J., Jr., Rehabilitation of streams receiving acid mine drainage, Bulletin 66, Virginia Poly. Inst. and State Univ., Virginia Water Resources Research Center, Blacksburg, Virginia, 1975.

Chapter 16 | Simulating Pollutographs and Loadographs

16-1 Introduction

A method is presented in this chapter that describes the quality of storm runoff as a function of time. A pollutograph (Fig. 16-1) describes pollutant *concentration* as a function of time, and a loadograph (Fig. 16-2) describes the cumulative pollutant *load* as a function of time. The correspondence of the pollutograph and loadograph to the storm runoff hydrograph is shown in Fig. 16-3. The hydrograph and pollutograph are established through actual measurement or by prediction using mathematical models that relate the system response to rainfall inputs. At any point in time following incipient runoff, the loadograph depicts the total pollutant load delivered by the stormwater up to, and including, that time. The loadograph is derived by convoluting the pollutograph with the hydrograph.

Engineers and planners dealing with nonpoint source pollution seek answers to the following questions:

1. What pollutants have been generated by human activities within the study area?

2. What are the rates of accumulation for these pollutants on different land use areas?

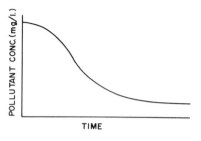

FIG. 16-1. A pollutograph.

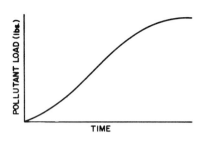

FIG. 16-2. A loadograph.

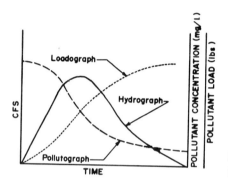

FIG. 16-3. A quantity/quality hydrograph illustrating the corresponding hydrograph, pollutograph, and loadograph.

3. How much of each pollutant is present in various runoff events, i.e., what is the stormwater quality for different levels of rainfall and runoff?

4. What is the impact of stormwater quality on receiving water bodies?

5. What will be the effect of land use changes and stormwater management practices on runoff quality?

The material presented in this chapter will aid the reader in answering questions 2, 3, and 5. Those pollutants generated by human activities within different land use areas were discussed in Chapter 15. The impact of stormwater quality on receiving streams is assessed with mathematical models of stream waste assimilation capacity that accept lateral inflow, the vector of nonpoint source pollution, and describe unsteady hydraulics.

Prior to the sections that deal with generating pollutographs and loadographs, the accumulation of pollutants and their subsequent removal by stormwater are discussed.

16-2   Accumulation of Pollutants

As discussed in Chapter 15, the nature and type of pollutants generated within a given area relate to the land use. The principal factors governing the rate of pollutant buildup on the land surface between storms are land use pattern and associated human activities, season of the year, and dustfall. Most studies designed to quantify the rates of pollutant accumulation have been conducted only in urban areas. Results from these studies, particularly those dealing with the accumulation on streets, are discussed in the following paragraphs.

Pollutants accumulate on streets from a variety of sources. Litter, tire and exhaust residue from automobiles, tree leaves, debris and matter washed into the streets from yards and other open areas, and dustfall contribute to the pollutant buildup. The pollutants commonly deposited on streets include BOD, COD, nutrients, metals, oil and grease, and solids.

The accumulation of pollutants on street surfaces results from two basic processes: deposition and removal [3]. It may be assumed that at any given site, the pollutants are deposited on streets at a constant rate for stable land use and similar seasonal conditions. As discussed in the next section, the rate of pollutant removal is proportional to the amount remaining on the street. Thus, the rate at which the pollutant load on the street changes can be expressed mathematically as [1]

$$dL/dt = C - kL = \text{rate of pollutant accumulation} \qquad (16\text{-}1)$$

where L is the pollutant loading on the street (lb/curb mile), C the constant rate of pollutant deposition (lb/curb mile/day), k the rate constant of pollutant removal, day^{-1} and t the time in days. Integration of Eq. (16-1) yields

$$L = (C/k)(1 - e^{-kt}) \qquad (16\text{-}2)$$

which describes the pollutant loading accumulated on the street as a function of the time elapsed since the last cleaning, either mechanical or by storm runoff.

The behavior of Eq. (16-2) is shown in Fig. 16-4. Initially, when t is small, the accumulation rate is rapid and almost constant, i.e., equal to the deposi-

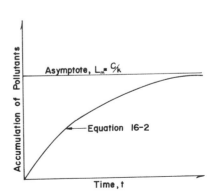

FIG. 16-4. Generalized plot of Eq. (16-2). Showing the asymptotic behavior as $t \to \infty$.

tion rate C. As time increases, the accumulation rate decreases and eventually, the accumulated load approaches the asymptote, C/k. This apparent upper limit may be visualized as a maximum possible pollutant loading L_m.

The functional relationship of Eq. (16-2) is observed in the solids accumulation curves for different land use shown in Fig. 16-5. These curves were determined from actual data from eight cities across the United States during a study of the water pollution aspects of the debris on streets [2]. This study was conducted by the URS Research Company for the US Environmental Protection Agency.

As part of the URS study, rainfall was simulated and allowed to flush the streets in the study areas for long durations of time. The amount of pollutants removed was plotted against time. These plots approached an asymptote which suggests that there are finite limits to the amount of material that could ever be washed from the street by rain. This correlates with the upper limit on accumulated pollutant loading.

Air currents, both natural and due to automobile traffic, which sweep the particles from the street, and natural decay are two reasons why an upper limit exists for the amount of pollutants that can accumulate on streets.

EXAMPLE 16.1   Determine the rate of pollutant deposition on a street which has the following flushing experiment results. After a 14-day dry

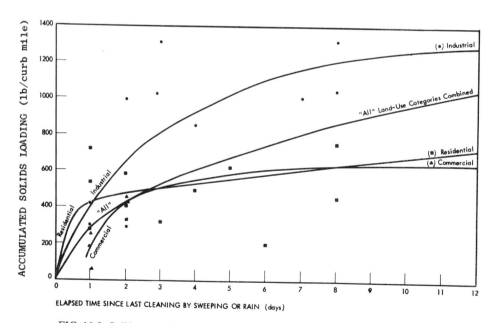

FIG. 16-5. Solids accumulation for different land use as a function of time, results of URS study for eight US cities. (After Sartor and Boyd [2].)

period a continuous 4-hr flushing at a simulated rainfall intensity of $\frac{1}{2}$ in./hr could remove only the equivalent of 250 lb of solids/curb mile. Two days later, another flushing experiment removed 138 lb.

Assume the 14-day dry period was long enough for the maximum loading to be reached and that no more solids would be removed by the flushing experiment. Set $L_m = 250$.

$$L = L_m(1 - e^{-kt}) \tag{16-2}$$

After two days,

$$L = 138 = 250(1 - e^{-k2}); \qquad k = 0.40$$

The constant deposition rate is given by

$$C/k = L_m, \qquad C = 0.40(250) = 100 \text{ lb/curb mile/day}$$

During the recent URS study mentioned above [2], it was found that a sizeable percentage of the pollution potential of street debris is contained in the very fine siltlike solids fraction (diameter $< 43\ \mu$). In this study, samples of street debris were collected in eight cities across the United States. The collection methods included simulating rainfall to ensure the collection of the very fine, soluble particles that may not be picked up by dry sweeping techniques. The very fine particles, though accounting only for 5.9% by weight of the total solids, contained approximately $\frac{1}{4}$ of the total oxygen demand, more than half of the heavy metals, and nearly $\frac{3}{4}$ of the total pesticides. Table 16-1 gives the breakdown of the individual pollutants associated with various particle size ranges.

TABLE 16-1 Fraction of pollutant associated with each particle size range, results from URS study of data from eight US cities (% by weight)[a]

Pollutant	Particle size (μ)					
	> 2000	840–2000	246–840	104–246	43–104	< 43
Total solids	22.4	7.6	24.6	27.8	9.7	5.9
Volatile solids	11.0	17.4	12.0	16.1	17.9	25.6
BOD$_5$	7.4	20.1	15.7	15.2	17.3	24.3
COD	2.4	4.5	13.0	12.4	45.0	22.7
Nitrates	8.6	6.5	7.9	16.7	28.4	31.9
Phosphates	0	0.9	6.9	6.4	29.6	56.2
Total pesticides	0	16.0	26.5	25.8	31.7	
Total heavy metals	16.3	17.5	14.9	23.5	27.8	
Lead	1.7	2.6	8.7	42.5	44.5	
Copper	22.5	20.0	16.5	19.0	22.0	
Mercury	16.4	28.8	16.4	19.2	19.2	

[a] After Sartor and Boyd [2].

The accumulation of pollutants on streets varies with land use and from city to city. Tables 16-2 and 16-3 show the average pollutant load and rate of pollutant accumulation on streets in the eight cities studied during the URS study. Tables 16-4 and 16-5 are comparable tables that present the data by land use category.

Certain trends are obvious from these data. The total solids data in Table 16-2 show there is a lower loading during the summer months than the winter months. This is anticipated since vastly larger amounts of dustfall are generated by the combustion for winter heating. One fact quite evident from Table 16-4 is that the rate of build up of pollutants on an urban watershed varies significantly with land use. These data show that industrial areas are much "dirtier" than residential areas.

TABLE 16-2 Average pollutant loads in eight US cities (lb/curb mile)[a]

City	Date of study	Curb miles	Total solids	Volatile solids	BOD$_5$	COD	Nitrates	Phosphates
San Jose	Dec. 1970	2300	910	66	16	310	—	0.70
Pheonix	Jan. 1971	2900	650	40	7	30	0.29	0.22
Milwaukee	Apr. 1971	3400	2700	180	12	48	0.052	0.27
Baltimore	May 1971	3900	1000	96	—	—	0.038	1.0
Seattle	July 1971	2600	460	29	5	17	0.027	0.49
Atlanta	June 1971	3500	430	18	2	13	0.024	0.26
Tulsa	June 1971	3600	330	19	14	30	0.12	0.54
Bucyrus	Apr. 1971	200	1400	150	3	29	0.12	0.25

[a] After Sartor and Boyd [2].

TABLE 16-3 Average rate of pollutant accumulation in eight US cities (lbs/curb mile/day)[a]

City	Date of study	Total solids	BOD$_5$	COD	Nitrate	Phosphate	Total heavy metals
San Jose	Dec. 1970	70	1.2	24	—	0.054	0.34
Pheonix	Jan. 1971	92	0.93	4.3	0.041	0.031	—
Milwaukee	Apr. 1971	2700	12.0	48	0.052	0.27	4.3
Baltimore	May 1971	260	—	—	0.0095	0.25	0.68
Seattle	July 1971	—	—	—	—	—	—
Atlanta	June 1971	220	0.95	6.5	0.012	0.13	0.21
Tulsa	June 1971	—	—	—	—	—	—
Bucyrus	Apr. 1971	690	1.4	25	0.060	0.12	—

[a] After Sartor and Boyd [2].

TABLE 16-4 Average pollutant loads by land use category, results of the URS study (lb/curb mile)[a]

	Land use category		
Pollutant	Residential	Industrial	Commercial
Total solids	1200	2800	360
Volatile solids	86	150	28
BOD_5	11	21	3
COD	25	100	7
Nitrates	0.06	0.18	0.18
Phosphates	1.1	3.4	0.3
Total heavy metals	0.58	0.76	0.18

[a] After Sartor and Boyd [2].

TABLE 16-5 Average rate of pollutant accumulation by land use category (lb/curb mile/day)[a]

	Land use category		
Pollutant	Residential	Industrial	Commercial
Total solids	590	1400	180
Volatile solids	44	77	14
BOD_5	3.6	7.2	0.99
COD	20	81	5.7
Nitrates	0.019	0.055	0.055
Phosphates	0.37	1.1	0.10
Total heavy metals	1.2	1.6	0.34

[a] After Sartor and Boyd [2].

In a later study that used the available data from 15 cities, the following conclusions regarding pollutant accumulation on streets were drawn [1]:

a. Loading rates are lowest in:
   1. Commercial areas, probably because they are swept frequently;
   2. The northwest;
   3. Areas with highest traffic, probably because the removal processes (primarily traffic generated winds) are more active;
   4. Tree-covered areas;
   5. Concrete surfaces.
b. BOD_5 concentrations are lowest in:
   1. Residential and heavy industrial areas;
   2. The southwest, probably reflecting the lack of lush vegetation relative to the East Coast;

    3.   Areas with moderate traffic;

    4.   Areas with landscaped buildings, probably reflecting better maintenance;

    5.   On asphalt road surfaces.

c. COD concentrations are lowest in:

    1.   Residential and heavy industry, whereas it is highest in commercial areas; the latter may be due to oil from many parked cars on the street;

    2.   Areas not significantly different climatologically;

    3.   Areas with moderate traffic;

    4.   Areas with landscaped buildings;

    5.   Areas not significantly different street surface types.

d. Ortho phosphate concentrations are lowest in:

    1.   Residential areas in an unexpected finding considering the widespread use of fertilizer on lawns;

    2.   The southwest (highest in southeast), probably reflecting the difference in vegetation or fertilizing practices;

    3.   Areas with moderate traffic;

    4.   Areas with no landscaping, probably because no fertilizer is used;

    5.   On asphalt surfaces.

e. Nitrate concentrations are lowest in:

    1.   Heavy industry areas;

    2.   Areas not significantly different climatologically;

    3.   Areas with no significantly different traffic densities;

    4.   Areas without landscaping;

    5.   Areas with no significantly different street surface types.

f. Lead concentrations are lowest in:

    1.   Heavy industry areas, low also in residential areas; this probably reflects low vehicular traffic;

    2.   The northeast (highest in the northwest); this may reflect the inhomogenity of sampling sites;

    3.   Areas with light and moderate traffic;

    4.   Areas with grass landscaping;

    5.   Concrete road surfaces.

g. Fecal coliform counts are lowest in:

    1.   Not significantly different land use categories; unexpected because generally it is thought that pet feces cause higher fecal coliform counts in residential areas;

    2.   The northwest;

    3.   Areas with heavy traffic;

    4.   Areas not significantly different in landscaping.

16-3   Removal of Pollutants

Rainfall on an impervious area must first wet the surface and fill the depression storage before generating any runoff. This first rain begins to dissolve the available water soluble pollutants. As rainfall continues, surface runoff begins and carries dissolved material with it. With increased flow and velocity, the suspended solids fraction is "picked up" and carried off the watershed. The settleable solids fraction is either suspended or "rolled along the surface" depending on the velocity. Runoff and associated pollutant removal are similar for a pervious watershed except that additional rainfall is lost to infiltration. This extra loss means less runoff from the pervious area and consequently less total pollutants removed by the stormwater, even though the concentrations may be as great at given discharges.

Not all the available pollutants are removed during a rainfall–runoff event. The percentage removed depends on the material itself, the land surface, the rainfall intensity, and especially the stormwater volume flowrate. Prediction of stormwater quality as a function of runoff intensity requires a mathematical model for the removal process that relates the amount of pollutant removed from the surface to varying rates of overland flow. In this section a mathematical formulation that predicts the sediment yield from sites available to erosion and a mathematical model that predicts the removal of pollutants on impervious surfaces, particularly streets, and a mathematical model for predicting urban pollutant concentrations as a function of storm characteristics are presented.

Sediment Yield

The major pollutant from construction, croplands, mining, and forest clear-cutting is sediment. With each of those activities the land has been denuded of vegetative cover and the soil disturbed. The first rainfall wets the soil causing it to swell and lose cohesion. Continued rainfall and surface runoff then cause *sheet* and *rill* erosion. The method of predicting sheet and rill erosion is based on the Universal Soil Loss Equation [4]. The soil loss equation is

$$A = RKLSCP \qquad (16\text{-}3)$$

where A is the computed soil loss per unit area, R the local rainfall factor, K the soil erodibility factor, L the soil slope length, and S the steepness of the slope. C and P are the cover-management and erosion control practice factors. These two factors are not used when predicting soil losses from sites where the soil has been disturbed.

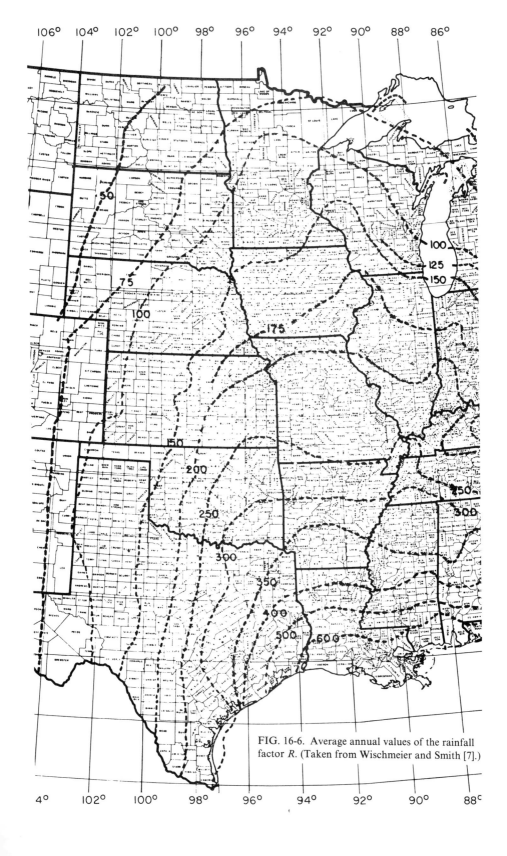

FIG. 16-6. Average annual values of the rainfall factor *R*. (Taken from Wischmeier and Smith [7].)

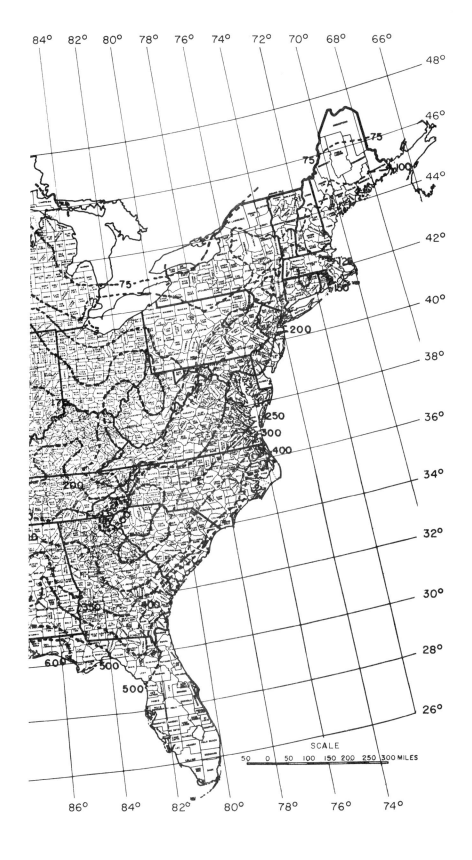

SCALE

50   0   50   100 150 200 250 300 MILES

In 1947, Musgrave [5] reported the results of analyses of soil loss measurements for some 40,000 storms occurring on small plots in the United States. His results indicated that the soil loss could be expressed in terms of an erodibility factor, cover factor, slope, length of slope, and the maximum 30-min rainfall intensity. Other studies, such as the one by Ellison [6], have demonstrated that erosion is dependent on the effects of raindrops. In 1965, Wischmeier and Smith [7] reported the Universal Soil Loss Equation which takes into account the influence of rainfall kinetic energy rather than just the amount of rainfall. The rainfall factor R for any storm is the product of the kinetic energy E of the rainfall and the maximum 30-min rainfall intensity I for that storm. That is, for any individual storm event, $R = EI$.

TABLE 16-6 Expected magnitudes of single storm erosion index values[a]

Location	Index values normally exceeded once in—					Location	Index values normally exceeded once in—				
	1 year	2 years	5 years	10 years	20 years		1 year	2 years	5 years	10 years	20 years
Alabama:						Maine:					
Birmingham	54	77	110	140	170	Caribou	14	20	28	36	44
Mobile	97	122	151	172	194	Portland	16	27	48	66	88
Montgomery	62	86	118	145	172	Skowhegan	18	27	40	51	63
Arkansas:						Maryland, Baltimore	41	59	86	109	133
Fort Smith	43	65	101	132	167	Massachusetts:					
Little Rock	41	69	115	158	211	Boston	17	27	43	57	73
Mountain Home	33	46	68	87	105	Washington	29	35	41	45	50
Texarkana	51	73	105	132	163	Michigan:					
California:						Alpena	14	21	32	41	50
Red Bluff	13	21	36	49	65	Detroit	21	31	45	56	68
San Luis Obispo	11	15	22	28	34	East Lansing	19	26	36	43	51
Colorado:						Grand Rapids	24	28	34	38	42
Akron	22	36	63	87	118	Minnesota:					
Pueblo	17	31	60	88	127	Duluth	21	34	53	72	93
Springfield	31	51	84	112	152	Fosston	17	26	39	51	63
Connecticut:						Minneapolis	25	35	51	65	78
Hartford	23	33	50	64	79	Rochester	41	58	85	105	129
New Haven	31	47	73	96	122	Springfield	24	37	60	80	102
District of Columbia	39	57	86	108	136	Mississippi:					
Florida:						Meridian	69	92	125	151	176
Apalachicola	87	124	180	224	272	Oxford	48	64	86	103	120
Jacksonville	92	123	166	201	236	Vicksburg	57	78	111	136	161
Miami	93	134	200	253	308	Missouri:					
Georgia:						Columbia	43	58	77	93	107
Atlanta	49	67	92	112	134	Kansas City	30	43	63	78	93
Augusta	34	50	74	94	118	McCredie	35	55	89	117	151
Columbus	61	81	108	131	152	Rolla	43	63	91	115	140
Macon	53	72	99	122	146	Springfield	37	51	70	87	102
Savannah	82	128	203	272	358	St. Joseph	45	62	86	106	126
Watkinsville	52	71	98	120	142	Montana:					
Illinois:						Great Falls	4	8	14	20	26
Cairo	39	63	101	135	173	Miles City	7	12	21	29	38
Chicago	33	49	77	101	129	Nebraska:					
Dixon Springs	39	56	82	105	130	Antioch	19	26	36	45	52
Moline	39	59	89	116	145	Lincoln	36	51	74	92	112
Rantoul	27	39	56	69	82	Lynch	26	37	54	67	82
Springfield	36	52	75	94	117	North Platte	25	38	59	78	99
Indiana:						Scribner	38	53	76	96	116
Evansville	26	38	56	71	86	Valentine	18	28	45	61	77
Fort Wayne	24	33	45	56	65	New Hampshire, Concord	18	27	45	62	79
Indianapolis	29	41	60	75	90	New Jersey:					
South Bend	26	41	65	86	111	Atlantic City	39	55	77	97	117
Terre Haute	42	57	78	96	113	Marlboro	39	57	85	111	136
Iowa:						Trenton	29	48	76	102	131
Burlington	37	48	62	72	81	New Mexico:					
Charles City	33	47	68	85	103	Albuquerque	4	6	11	15	21
Clarinda	35	48	66	79	94	Roswell	10	21	34	45	53
Des Moines	31	45	67	86	105	New York:					
Dubuque	43	63	91	114	140	Albany	18	26	38	47	56
Rockwell City	31	49	76	101	129	Binghamton	16	24	36	47	58
Sioux City	40	58	84	105	131	Buffalo	15	23	36	49	61
Kansas:						Marcellus	16	24	38	49	62
Burlingame	37	51	69	83	100	Rochester	13	22	38	54	75
Coffeyville	47	69	101	128	159	Salamanca	15	21	32	40	49
Concordia	33	53	86	116	154	Syracuse	15	24	38	51	65
Dodge City	31	47	76	97	124	North Carolina:					
Goodland	26	37	53	67	80	Asheville	28	40	58	72	87
Hays	35	51	76	97	121	Charlotte	41	63	100	131	164
Wichita	41	61	93	121	150	Greensboro	37	51	74	92	113
Kentucky:						Raleigh	53	77	110	137	168
Lexington	28	46	80	114	151	Wilmington	59	87	129	167	206
Louisville	31	43	59	72	85	North Dakota:					
Middlesboro	28	38	52	63	73	Devils Lake	19	27	39	49	59
Louisiana:						Fargo	20	31	54	77	103
New Orleans	104	149	214	270	330	Williston	11	16	25	33	41
Shreveport	55	73	99	121	141	Ohio:					
						Cincinnati	27	38	48	59	69
						Cleveland	22	35	55	71	86
						Columbiana	20	26	35	41	48
						Columbus	27	40	60	77	94
						Coshocton	27	45	77	108	143
						Dayton	21	30	44	57	70
						Toledo	16	26	42	57	74

Figure 16-6 gives average annual values of R in terms of isoerodents for that portion of the United States east of the Rocky Mountains. Isoerodents in the mountainous states west of the 104th meridian were not included because of the sporadic rainfall pattern of the mountains. When the soil loss equation is used to estimate average annual soil loss for a given locale, the *average annual* value of R is obtained by interpolation between adjacent isoerodent lines. To approximate the amount of soil loss from a single storm that probably will be exceeded once in 1, 2, 5, 10, and 20 yr, the appropriate R values in Table 16-6 should be used.

The soil erodibility factor K has been determined experimentally for the major soil groups. For any particular soil, K represents the rate of erosion per unit of erosion index from unit plots on that soil. The erosion index

Location	Index values normally exceeded once in—				
	1 year	2 years	5 years	10 years	20 years
Oklahoma:					
Ardmore	46	71	107	141	179
Cherokee	44	59	80	97	113
Guthrie	47	70	105	134	163
McAlester	54	82	127	165	209
Tulsa	47	69	100	127	154
Oregon, Portland	6	9	13	15	18
Pennsylvania:					
Franklin	17	24	35	45	54
Harrisburg	19	25	35	43	51
Philadelphia	28	39	55	69	81
Pittsburgh	23	32	45	57	67
Reading	28	39	55	68	81
Scranton	23	32	44	53	63
Puerto Rico, San Juan	57	87	131	169	216
Rhode Island, Providence	23	34	52	68	83
South Carolina:					
Charleston	74	106	154	196	240
Clemson	51	73	106	133	163
Columbia	41	59	85	106	132
Greenville	44	65	96	124	153
South Dakota:					
Aberdeen	23	35	55	73	92
Huron	19	27	40	50	61
Isabel	15	24	38	52	67
Rapid City	12	20	34	48	64
Tennessee:					
Chattanooga	34	49	72	93	114
Knoxville	25	41	68	93	122
Memphis	43	55	70	82	91
Nashville	35	49	68	83	99
Texas:					
Abilene	31	49	79	103	138
Amarillo	27	47	80	112	150
Austin	51	80	125	169	218
Brownsville	73	113	181	245	312
Corpus Christi	57	79	114	146	171
Dallas	53	82	126	166	213
Del Rio	44	67	108	144	182
El Paso	6	9	15	19	24
Houston	82	127	208	275	359
Lubbock	17	29	53	77	103
Midland	23	35	52	69	85
Nacogdoches	77	103	138	164	194
San Antonio	57	82	122	155	193
Temple	53	78	123	162	206
Victoria	59	83	116	146	178
Wichita Falls	47	63	86	106	123
Vermont, Burlington	15	22	35	47	58
Virginia:					
Blacksburg	23	31	41	48	56
Lynchburg	31	45	66	83	103
Richmond	46	63	86	102	125
Roanoke	23	33	48	61	73
Washington, Spokane	3	4	7	8	11
West Virginia:					
Elkins	23	31	42	51	60
Huntington	18	29	49	69	89
Parkersburg	20	31	46	61	76
Wisconsin:					
Green Bay	18	26	38	49	59
LaCrosse	46	67	99	125	154
Madison	29	42	61	77	95
Milwaukee	25	35	50	62	74
Rice Lake	29	45	70	92	119
Wyoming:					
Casper	4	7	9	11	14
Cheyenne	9	14	21	27	34

[a] After Wischmeier and Smith [7].

(*EI*) is the rainfall factor discussed above. In the erosion determination experiments, a "unit" plot is 72.6 ft long, with a uniform lengthwise slope of 9%, in continuous fallow tilled up and down the slope. Values of *K* determined for 23 major soils on which erosion plot studies were conducted are given in Table 16-7.

The rate of soil erosion by stormwater is greatly affected by both the slope of a site and the length of this slope. The combined effects are expressed by the factor *LS* which is the expected ratio of soil loss per unit area on a study site to the corresponding loss from a "unit" plot. This ratio may be taken directly from the slope-effect chart, Fig. 16-7. Values of *LS* for slope percentage not shown on the chart may be computed by solving the following equation:

$$LS = \sqrt{\lambda}(0.0076 + 0.0053s + 0.0076s^2) \qquad (16\text{-}4)$$

TABLE 16-7   Computed *K* values for soils on erosion research stations[a]

Soil	Source of data	Computed *K*
Dunkirk silt loam	Geneva, New York	0.69
Keene silt loam	Zanesville, Ohio	0.48
Shelby loam	Bethany, Missouri	0.41
Lodi loam	Blacksburg, Virginia	0.39
Fayette silt loam	LaCrosse, Wisconsin	0.38
Cecil sandy clay loam	Watkinsville, Georgia	0.36
Marshall silt loam	Clarinda, Iowa	0.33
Ida silt loam	Castana, Iowa	0.33
Mansic clay loam	Hays, Kansas	0.32
Hagerstown silty clay loam	State College, Pennsylvania	0.31
Austin clay	Temple, Texas	0.29
Mexico silt loam	McCredie, Missouri	0.28
Honeoye silt loam	Marcellus, New York	0.28
Cecil sandy loam	Clemson, South Carolina	0.28
Ontario loam	Geneva, New York	0.27
Cecil clay loam	Watkinsville, Georgia	0.26
Boswell fine sandy loam	Tyler, Texas	0.25
Cecil sandy loam	Watkinsville, Georgia	0.23
Zaneis fine sandy loam	Guthrie, Oklahoma	0.22
Tifton loamy sand	Tifton, Georgia	0.10
Freehold loamy sand	Marlboro, New Jersey	0.08
Bath flaggy silt loam with surface stones >2 in. removed	Arnot, New York	0.05
Albia gravelly loam	Beemerville, New Jersey	0.03

[a] After Wischmeier and Smith [7].

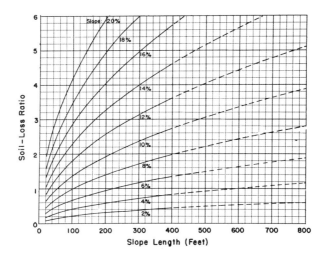

FIG. 16-7. Slope effect chart (topographic factor, *LS*). (After Wischmeier and Smith [7].)

where λ is the length of the slope in feet, and s the grade of the slope expressed in percent.

Use of the soil loss equation is demonstrated in the following example problem.

EXAMPLE 16.2   Given a 20-acre construction site in Knoxville, Tennessee, with an average slope of 10% and a slope length of 400 ft. The local SCS office has classified the soil as similar in properties to the Hagerstown silty clay loam. Estimate

(a)   the expected soil loss from the site during a 12-month period, and

(b)   the soil loss from a single storm that may be exceeded once in 10 yr.

From Table 16-7 the erodibility factor K is found to be 0.31. The topographic factor LS is found on Fig. 16-7 by following up the slope length = 400 line to the intersection with slope = 10%. Projecting across to the soil-loss ratio axis, $LS = 2.73$.

(a)   The annual rainfall factor is determined using Fig. 16-6. Knoxville is in Knox County, which lies between the 175 and 200 isoerodents. Interpolating linearly, $R = 190$.

The estimated annual soil loss is

$$A = RKLS = 190 \times 0.31 \times 2.73 = 160.8 \text{ tons/acre.}$$

For the 20 acres, $160.8 \times 20 = 3216$ tons.

(b)   From Table 16-6, for a single storm that may be exceeded once in 10 yr, $R = 93$.

The estimated soil loss for this storm is

$$A = 93 \times 0.31 \times 2.73 = 78.7 \text{ tons/acre}$$

From 20 acres, $78.7 \times 20 = 1574$ tons.

Removal of Pollutants from Impervious Areas

The removal of pollutants from impervious surfaces is a direct function of the *total volume of storm runoff* [1]. Table 16-8 shows representative percentages of pollutant removed from streets by various combinations of runoff rate and duration. Note that a runoff rate of 0.5 in./hr which lasts for 1 hr will remove the same percentages of pollutants as a runoff rate of 1.0 in./hr that lasts for $\frac{1}{2}$ hr. In both cases, $\frac{1}{2}$ in. of total runoff occurs.

A deterministic model is derived that computes the amount of pollutant washed off the total urban watershed during a storm. This formulation is basically the same as the overland flow quality model that is a part of the EPA Stormwater Management model.

The following assumptions are made as the basis for the mathematical derivation [8]:

1. No pollutants decay due to chemical changes or biological degradation during the runoff process.

2. The amounts of pollutants percolating into the soil by infiltration are neglected.

3. The rate of removal of pollutants by runoff water is assumed to be proportional to the amount of pollutant remaining, and to the runoff intensity.

TABLE 16-8 Percent of pollutants removed from street surfaces by runoff rate/duration[a]

Runoff rate (in./hr)	Runoff duration (hr)							
	0.25	0.5	1.0	2.0	3.0	4.0	5.0	6.0
0.1	10.9	20.5	36.0	60.1	74.9	84.1	90.0	90.0
0.2	20.5	36.9	60.1	84.1	>90.0	>90.0	>90.0	
0.3	29.1	49.8	74.9	>90.0				
0.4	36.9	60.1	84.1					
0.5	43.7	68.3	90.0					
0.6	49.8	74.8	>90.0					
0.7	55.3	80.0						
0.8	60.1	84.1						
0.9	64.5	87.4						
1.0	68.3	90.0						

[a] After US Environmental Protection Agency [1].

4. Because the distribution of air pollutants in the atmosphere vary significantly in space and time, this source is not evaluated separately but is considered to be in the overall runoff constituents.

The assumption that the amount of pollutant removed at any time t is proportional to the amount remaining is expressed mathematically by

$$dP/dt = -kP \qquad (16\text{-}5)$$

where P is the amount of pollutant remaining at time t, k the velocity of pollutant removal by the stormwater, and t the time since the start of the storm runoff. Since the runoff rate Q affects the rate of pollutant removal (assumption 3), k must be functionally dependent upon Q. The rationale for determining k was discussed by Roesner [9].

> Given two identical watersheds except for their area size, for the same rainfall rate r on both watersheds a higher runoff rate would occur from the larger watershed. This area effect can be eliminated by dividing the runoff Q by the *impervious* area of the watershed. The impervious area is used because only a negligible amount of the runoff comes from the pervious area. Since cubic feet per second per acre are equivalent to inches per hour, we can say that k is functionally dependent on the runoff rate R from the impervious area, where R is in in./hr. Finally, assuming that k is directly proportional to R and that a uniform rainfall of $\frac{1}{2}$ in./hr would wash away 90% of the pollutant in 1 hr, we can say that $k = 4.6R$.

This value of k for the rate of pollutant removal from street surfaces was verified in a study of the pollution contributed by streets by Sartor and Boyd [5].

Substituting the expression for k into Eq. (16-5), holding R constant, and integrating with respect to time, yields the following equation that predicts the amount of pollutant remaining on the impervious area after time t of storm runoff:

$$P = P_0 \exp(-4.6Rt) \qquad (16\text{-}6)$$

where P_0 is the initial pollutant loading on the impervious surface, and R the rate of overland flow in inches per hour from the impervious area. The behavior of this equation is shown in Fig. 16-8. At a constant runoff rate, stormwater removes pollutants from the watershed according to the exponential decay. The rate of this decay varies directly with the rate of flow.

One limitation of the above relationship is that it is valid only for a steady (constant) runoff rate. Unfortunately, the runoff during the rising and falling limbs of the hydrograph is unsteady. Integrating Eq. (16-5) over a

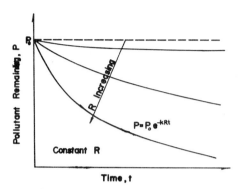

FIG. 16-8. Behavior of the overland flow quality model under steady runoff. (After Roesner [9].)

finite time interval of length Δt (during which R is held constant) rather than over continuous time dt gives

$$P(t + \Delta t) = P(t) \exp(-4.6R \, \Delta t) \tag{16-7}$$

This equation predicts the amount of pollutant remaining in the watershed at the end of Δt, $P(t + \Delta t)$, in terms of the amount present at the beginning of Δt, $P(t)$, where $P(t) < P(t + \Delta t)$. By choosing Δt small enough and using the average runoff rate during each Δt interval $\bar{R} = \frac{1}{2}[R(t + \Delta t) + R(t)]$, a very good approximation is made to the pollutant removal process during the times of unsteady flow. $R(t)$ is the runoff rate measured in inches per hour at time t. Selection of a Δt that is overly large may introduce significant error, particularly if the runoff rate is highly variable. Figure 16-9 illustrates the application of Eq. (16-7) to predict the removal process during the times of unsteady flow.

For the purpose of developing a pollutograph, the amount of pollutant removed from the study area by the storm runoff is required. With loadings typically expressed in terms of pounds per curb mile, the removal is based

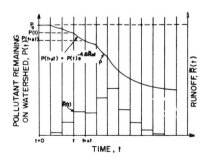

FIG. 16-9. Pollutant remaining on watershed during stormwater runoff event.

upon street runoff. This amount is divided by the runoff from the impervious area to establish the predicted concentration.

The amount of pollutant removed from the street surface after time t is the amount remaining subtracted from the initial loading:

$$\text{amount removed} = P_0 - P = P_0(1 - \exp[-4.6Rt]) \qquad (16\text{-}8)$$

For the discrete case, the amount removed during the time interval Δt is given by

$$P(t) - P(t + \Delta t) = \Delta P = P(t)(1 - \exp[-4.6\bar{R}\,\Delta t]) \qquad (16\text{-}9)$$

Equation (16-9) is applied to an example problem in the next section that demonstrates the calculation of a pollutograph and a loadograph.

A matter of concern that arises when predicting the pollutant concentration in stormwater with Eqs. (16-8) and (16-9) involves the use of impervious area runoff only. By neglecting the runoff contribution from pervious areas, either the potential for further dilution is overlooked and a "worst case" condition is predicted, or additional pollutant sources are precluded. The value of k' ($k' = 4.6$) in Eqs. (16-8) and (16-9) is the value established for pollutant removal from streets, and represents the largest value for k'. A smaller k' ($k' < 4.6$) suggests pollutant contributions both from pervious areas and other impervious areas in addition to the streets. In this case, the removal of a given percentage of the total pollutant load accumulated on the study area requires a larger total volume of storm runoff. In a study of an urban watershed in Knoxville, Tennessee, Barkdoll [10] evaluated the k' values for various pollutants. The watershed for this study was a 1.6 sq. mile area that was estimated to be 45% impervious, including 22.4 miles of streets and roads. All k' values were significantly less than 4.6, ranging from $k' = 0.16$ for calcium to $k' = 2.0$ for mercury. It was concluded from this study that values of k' are (1) site specific, and (2) pollutant specific. Obviously, the runoff volume and pollutant concentration from pervious areas should be considered, particularly in those study areas with steep slopes, frequent rainfall events, and relatively impervious soils.

Regression Models

When sufficient data exist, a model that relates storm characteristics to runoff water quality may be developed through regression analysis. Least squares regression (see Chapters 9 and 14) determines the "optimal" parameter values for a model of predetermined form by minimizing the sum of the squares of the difference between the observed values of the dependent variable and the values predicted from the data by the regression model.

This modeling approach is considered parametric if the model structure is *logical* to the process being modeled and the process is stable; e.g., the runoff process is stable during a study period as long as no significant land use changes are made. If there is a strong element of chance involved, or if the model structure is "black box," the model is stochastic. Since the parameter values are data dependent a regression model is site specific.

Two popular model forms are

$$Y = a + bX_1 + cX_2 + dX_3 + eX_4 \qquad (16\text{-}10)$$

$$Y = aX_1{}^b X_2{}^c X_3{}^d X_4{}^e \qquad (16\text{-}11)$$

where Y is the dependent variable, X_1, X_2, X_3, X_4 are independent variables, and a, b, c, d, e are parameters to be determined based on the data by a least squares analysis.

These equations were written in terms of four independent variables only for illustrative purposes. The number of independent variables included in a model should be enough to adequately describe the process, but not so many as to render the equation meaningless and untenable. Equation (16-10) is a linear model, whose parameters are determined by linear least squares (Chapter 9). The second equation, (16-11), is a nonlinear model. The parameter values for this equation are determined by nonlinear least squares (Chapter 9) or by linear least squares following the transformation

$$\log Y = \log a + b \log X_1 + c \log X_2 + d \log X_3 + e \log X_4 \quad (16\text{-}12)$$

which expresses the nonlinear equation in linear form.

An application of least squares regression to stormwater quality modeling was demonstrated by Colston in his study of urban storm runoff at Durham, North Carolina [11]. As part of the study, regression equations were developed that described within-storm variations for 19 quality variables in terms of storm characteristics. A nonlinear model of the form of Eq. (16-11) was used. The independent variables were rate of runoff (CFS), time from the start of a storm (TFSS) in hours, time from last storm (TFLS) in hours, and time from last peak (TFLP) in hours. A stepwise regression using the data from 36 storm events found that the rate of discharge (CFS) and the time from the storm start (TFSS) were the two most significant variables. Only a modest gain in the coefficient of determination r^2 was observed when including the other two time variables. For this reason, Colston [11] decided to limit the regression equations to CFS and TFSS for regression simplicity. The equations determined for the 19 quality variables are shown in Table 16-9. These equations are specific to the Third Fork Creek in Durham, North Carolina.

It is significant that the final model form relates the stormwater pollutant load to a product of the rate of discharge and the time since the start of the

TABLE 16-9   Equations describing urban runoff pollutant flux (lb/min) for Durham, North Carolina as a function of discharge rate (cfs) and time from storm start (TFSS) (hr)[a]

Equation	r^2
$COD = 0.51\ CFS^{1.11}\ TFSS^{-0.28}$	0.90
$TOC = 0.16\ CFS^{1.0}\ TFSS^{-0.28}$	0.84
Total solids $= 3.35\ CFS^{1.14}\ TFSS^{-0.18}$	0.85
Volatile solids $= 0.58\ CFS^{1.09}\ TFSS^{-0.11}$	0.92
Suspended solids $= 1.89\ CFS^{1.23}\ TFSS^{-0.16}$	0.76
Volatile suspended solids $= 0.25\ CFS^{1.18}\ TFSS^{-0.17}$	0.83
Kjeldahl nitrogen $= 0.0032\ CFS^{0.87}\ TFSS^{-0.29}$	0.73
Total phosphorus as $P = 0.003\ CFS^{1.03}\ TFSS^{-0.29}$	0.92
Aluminum $= 0.0443\ CFS^{1.05}\ TFSS^{-0.15}$	0.89
Calcium $= 0.045\ CFS^{0.60}\ TFSS^{-0.09}$	0.82
Cobalt $= 0.0003\ CFS^{1.18}\ TFSS^{+0.13}$	0.92
Chromium $= 0.0008\ CFS^{0.96}\ TFSS^{+0.06}$	0.89
Copper $= 0.00035\ CFS^{1.10}\ TFSS^{+0.08}$	0.94
Iron $= 0.0238\ CFS^{1.24}\ TFSS^{-0.18}$	0.87
Lead $= 0.0013\ CFS^{1.125}\ TFSS^{-0.29}$	0.83
Magnesium $= 0.0434\ CFS^{0.98}\ TFSS^{-0.16}$	0.94
Manganese $= 0.0023\ CFS^{1.11}\ TFSS^{-0.27}$	0.94
Nickel $= 0.0005\ CFS^{1.03}\ TFSS^{+0.01}$	0.94
Zinc $= 0.0011\ CFS^{1.10}\ TFSS^{-0.22}$	0.89

[a] After Colston [11].

storm, and did not include the other two time variables. This agrees with the statement in conjunction with the derivation of the deterministic pollutant removal model that the "removal of pollutants is a direct function of the total volume of runoff." The time since the last storm was not important for two reasons: (1) the frequency of storm events, and (2) the fact that a major portion of the pollutants present at the outset of a storm have accumulated on the basin in the first one or two days following the last storm. The latter point was observed by Sartor and Boyd [2] and is shown in Fig. 16-5. Since the time between most storms exceeds one or two days, an equivalent amount of pollutants will have accumulated prior to most runoff events. The time from the last peak was not significant since, again, it is the volume of runoff that is important in the removal of pollutants by stormwater.

16-4   Simulating Pollutographs and Loadographs

Pollutographs and loadographs are generated by mathematical models that predict pollutant concentrations for the different discharges in a storm runoff event. Therefore, varying concentrations relate to the hydrograph and can be expressed as a function of time. Those models that include the pollutant removal process, such as the overland flow quality model discussed in the last section, allow for different concentrations at different times within the runoff event for flowrates of the same intensity. Other models that relate pollutant concentrations only to discharge will predict the same concentration for equivalent flowrates, irrespective of the position of the flows relative to the beginning of the storm runoff. Depending on the modeler's objectives and the data available, a wide range of choice of stormwater runoff models exist. Some models have the capability of simulating both runoff quantity and quality while other models can predict only quantity and require a linkage with the modeler's own choice of quality models. The following examples demonstrate the application of stormwater runoff models in simulating stormwater quality. Each runoff model was discussed in this text.

Urban Stormwater Quality—the EPA Stormwater Management Model

The Stormwater Management model [12] is a deterministic model that takes rainfall and basin characteristics as input and calculates stormwater quantity and quality. The model components that generate runoff hydrographs and route the stormwater through city storm sewers were discussed in Chapter 6. Simulation of the stormwater quality involves the use of the overland flow quality model derived in Section 16-3. Before discussing an application of the total model, an example problem will be worked that demonstrates the generation of a pollutograph and loadograph by the component runoff quality model.

EXAMPLE 16.3   Given a 320-acre urban watershed that is 55% impervious and has 30 miles of streets which comprise 35% of the total area. Using the hydrograph in Fig. 16-10, develop a corresponding pollutograph and loadograph based on a total solids loading of 500 lb/curb mile.

Since the loading is given in terms of streets rather than the total impervious area, two hydrographs are required. The street runoff hydrograph is used in generating the amount of pollutants removed which is then diluted by the total impervious area runoff. If the loading had been given for the total impervious area, only the impervious area hydrograph would be needed.
The general pollutograph equation is

$$\Delta P = P(t)(1 - \exp[-4.6\bar{R}\,\Delta t]) \qquad (16\text{-}9)$$

where ΔP is the amount of pollutants removed during the time interval Δt, $P(t)$ the amount of pollutants remaining at the beginning of the time interval Δt, $\bar{R}$ the average runoff during the time interval Δt (in./hr), and Δt the

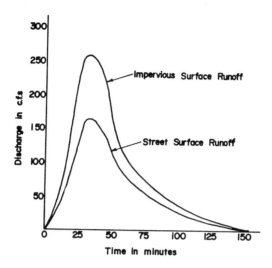

FIG. 16-10. Hydrographs for Example 16.3.

length of the time interval (hr). The initial pollutant loading is given by

$$(\text{miles of street}) \times (2\ \text{curbs/street}) \times (\text{pollutant load lb/curb-mile})$$
$$= 30 \times 2 \times 500 = 30{,}000\ \text{lb}.$$

The worksheet for the solution of this problem is given as Table 16-10. The computed pollutograph and loadograph are shown in Fig. 16-11.

Application of the EPA Stormwater Management model was demonstrated on the Bloody Run drainage basin in Cincinnati, Ohio [12]. The model was used to predict the quantity and quality of combined sewer flow

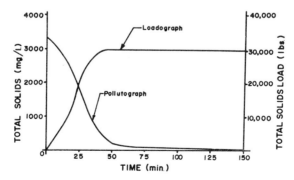

FIG. 16-11. Pollutograph and loadograph developed in Example 16-3.

TABLE 16-10 Pollutograph and loadograph worksheet[a]

	Street surface runoff							Impervious area runoff				
Δt (min)	$R(t)$ (CFS)	$R(t+\Delta t)$ (CFS)	$\bar{R}$ (CFS)	$\bar{R}$ (in./hr)	$P(t)$ (lb)	ΔP (lb)	$\sum \Delta P$ (lb)	$R(t)$ (CFS)	$R(t+\Delta t)$ (CFS)	$\bar{R}$ (CFS)	Volume (ft³)	Total solids concentration (mg/l)
0–12.5	0	45	22.5	0.201	30,000	5256	5256	0	80	40	30,000	2806
12.5–25	45	137	91	0.813	24,744	13,391	18,647	80	230	155	116,250	1845
25–37.5	137	160	148.5	1.326	11,353	8167	26,814	230	253	241.5	181,125	722
37.5–50	160	113	136.5	1.219	3186	2195	29,009	253	165	209	156,750	224
50–62.5	113	75	94	0.839	991	547	29,556	165	104	134.5	100,875	87
62.5–75	75	55	65	0.580	444	189	29,745	104	72	88	66,000	46
75–87.5	55	37	46	0.411	255	83	29,828	72	55	63.5	47,625	28
87.5–100	37	27	32	0.286	172	41	29,869	55	37	46	34,500	19
100–112.5	27	18	22.5	0.201	131	23	29,892	37	25	31	23,250	16
112.5–125	18	12	15	0.134	108	13	29,905	25	15	20	15,000	14
125–137.5	12	7	9.5	0.085	95	7	29,912	15	10	12.5	9375	12
137.5–150	7	0	3.5	0.031	88	2	29,914	10	0	5	3750	8

[a]The following equations have been used in preparing the table:

$$\bar{R}(\text{CFS}) = \tfrac{1}{2}[R(t) + R(t + \Delta t)]; \qquad \Delta P(\text{lb}) = P(t)[1 - \exp(-4.6\bar{R}\,\Delta t/60)]; \qquad \bar{R}(\text{in./hr}) = \bar{R}(\text{CFS})/\text{area of streets};$$

$$\text{volume (ft}^3) = \bar{R} \times (\Delta t) \times 60; \qquad \text{conc (mg/l)} = \Delta P(1\ \text{ft}^3/28,316\ \text{l})(453,592.37\ \text{mg/lb})/\text{runoff vol. from impervious area.}$$

during storm conditions. Several points in the collection system were monitored simultaneously as a test of the flow and quality routing efficiency of
the model.

The test area, shown in Fig. 16-12, is composed of 2380 areas of hilly
land. At the time of the study, 55% of the area was residential, 17% commercial, 5% industrial, and 22% open land. The drainage basin has two main
valleys running east to west. Most of the commercial and industrial sections
are in the valleys, and the residential areas are on the ridges. The overall
areal population density is 11 persons/acre.

The basin was divided into 38 catchments on the basis of land use,
population density, and topography. The data required for each of these
subareas were rainfall, average overland slope, percent imperviousness, and
length and slope of sewer. Using these inputs the model was run with computations being made at 5-min intervals.

Initially, the model was verified for dry flow conditions. Characteristic
flows and qualities were input to the model and routed through the sewer
system and the results are given in Table 16-11. The point of discharge from
the basin is sampling point number six. At this station the predicted and
observed flows correspond well, but noticeable differences exist between the
observed and predicted BOD and SS. The error between the observed and
predicted is <3% for flow, 25% for BOD and 11% for SS.

Figure 16-13 shows the model results for the combined sewer flow at
the mouth of the basin for the storm of April 1, 1970. The time of peak is

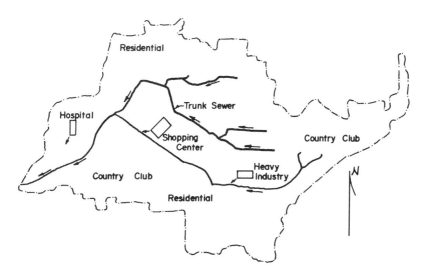

FIG. 16-12. Bloody Run drainage basin, Cincinnati, Ohio. (After EPA [12].)

TABLE 16-11   Cincinnati dry weather flow results[a]

Sampling location	Flow (cfs)		BOD (mg/l)		SS (mg/l)	
	Reported[b]	Computed	Reported[b]	Computed	Reported[b]	Computed
1	0.93	0.90	360	403	224	206
2	0.54	0.50	350		230	
3	1.45	2.12	1160		236	
4	15.50	12.58	618	529	265	226
5	0.54	0.80	292		181	
6	13.94	13.61	412	517	252	224

[a] After US Environmental Protection Agency [12].

[b] Reported values are averages of approximately ten grab samples each over a two-week period.

the same for both the predicted and observed hydrographs, but the volume of flow for the computed hydrograph is well below the measured hydrograph. One logical explanation is the variation in rainfall. The rainfall was distributed over the basin on the measurements at two rain gauges that are both outside of the basin. The variance in the observed and predicted pollutographs is anticipated since the generation of a pollutograph is tied so closely to the runoff hydrograph.

Simulating Mineral Load—TVA Daily Flow Model

The TVA daily flow model [13] has been used by Betson and McMaster [14] to simulate mineral loads in the Tennessee Valley. The linkage between flow and quality was achieved by use of the rating

$$C = a(Q/DA)^b \qquad (16\text{-}13)$$

where C is the concentration of the mineral constituent in milligrams per liter, Q the streamflow in cubic feet per second, DA is drainage area in sq miles, and a and b are empirically determined coefficients.

The two coefficients in Eq. (16-13) were related to land use, soils, quality and quantity of rainfall, and geologic factors by the linear model.

$$a,b = c_1F + c_2C + c_3S + c_4I + c_5U \qquad (16\text{-}14)$$

where C_i is a regression coefficient, F the fraction of watershed in forest, C the drainage area fraction over carbonate rock, S drainage area fraction over shale–sandstone rock, I the drainage area fraction over igneous rock, and U the drainage area fraction over consolidated rock. The four geologic variables allocate the drainage area among the rock types and must sum to unity.

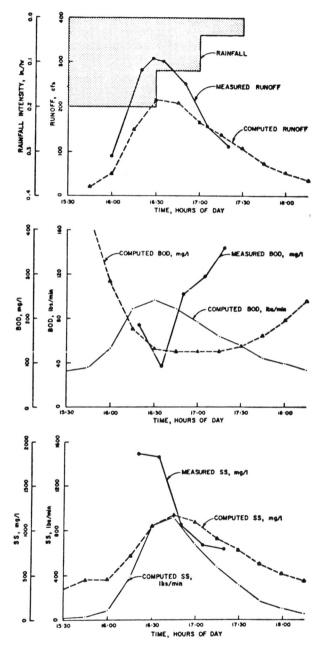

FIG. 16-13. Cincinnati combined sewer overflow results—storm of April 1, 1970. (After EPA [12].)

TABLE 16-12 Nonpoint source water quality model watersheds[a]

Code	Watershed	Forest	Independent Variables			
			C	S	I	U
	Calibration watersheds:					
	Mississippi Embayment:					
BRNC	Beech River near Chesterfield, Tenn.-New Channel	0.45	0.0	0.0	0.0	1.00
BR-L	Beech River near Lexington, Tenn.	0.37	0.0	0.0	0.0	1.00
BS-B	Big Sandy River at Bruceton, Tenn.	0.50	0.0	0.0	0.0	1.00
CCSH	Cane Creek near Shady Hill, Tenn.	0.16	0.0	0.0	0.0	1.00
	Highland Rim and Nashville Basin:					
BNCC	Big Nance Creek at Courtland, Ala.	0.40	0.48	0.52	0.0	0.0
BF62	Boiling Fork Creek above Winchester, Tenn.	0.48	0.76	0.24	0.0	0.0
CC-H	Coon Creek near Hohenwald, Tenn.	0.83	0.16	0.84	0.0	0.0
CCFA	Cypress Creek near Florence, Ala.	0.39	0.28	0.30	0.0	0.42
FR-M	Flint River near Maysville, Ala.	0.29	0.42	0.58	0.0	0.0
HCLT	Holland Creek near Loweryville, Tenn.	0.89	0.0	0.31	0.0	0.69
HC-S	Horse Creek near Savannah, Tenn.	0.80	0.03	0.34	0.0	0.63
LC-A	Limestone Creek near Athens, Ala.	0.27	0.21	0.79	0.0	0.0
PR-W	Paint Rock River near Woodville, Ala.	0.82	0.35	0.65	0.0	0.0
RC-C	Richland Creek near Cornersville, Tenn.	0.28	0.81	0.19	0.0	0.0
RC-P	Richland Creek near Pulaski, Tenn.	0.32	0.63	0.37	0.0	0.0
SCGS	Sugar Creek near Good Springs, Ala.	0.26	0.13	0.87	0.0	0.0
YC-C	Yokley Creek near Campbellsville, Tenn.	0.46	0.54	0.46	0.0	0.0

Cumberland Plateau:

Code	Station					
CCFG	Chattanooga Creek near Flintstone, Ga.	0.80	0.40	0.60	0.0	0.0
COCD	Crab Orchard Creek near Deermont, Tenn.	0.81	0.0	1.00	0.0	0.0
DC-H	Daddy's Creek near Hebbertsburg, Tenn.	0.76	0.02	0.98	0.0	0.0
ER-O	Emory River at Oakdale, Tenn.	0.87	0.0	1.00	0.0	0.0
ER-W	Emory River near Wartburg, Tenn.	0.95	0.0	1.00	0.0	0.0
NFPR	N. Fork Powell River at Pennington Gap, Va.	0.82	0.0	1.00	0.0	0.0
OR-L	Obed River near Lansing, Tenn.	0.84	0.01	0.99	0.0	0.0
SFPR	S. Fork Powell River at Big Stone Gap, Va.	0.47	0.19	0.81	0.0	0.0
SRCS	Sequatchie River near College Station, Tenn.	0.56	0.57	0.43	0.0	0.0
SR-W	Sequatchie River near Whitwell, Tenn.	0.67	0.43	0.57	0.0	0.0
TC-G	Town Creek near Geraldine, Ala.	0.59	0.14	0.86	0.0	0.0
WCGA	Whites Creek at Glen Alice, Tenn.	0.83	0.09	0.90	0.01	0.0

Valley and Ridge:

Code	Station					
BC-K	Beech Creek at Kepler, Tenn.	0.79	0.0	1.00	0.0	0.0
BC-R	Big Creek near Rogersville, Tenn.	0.54	0.30	0.70	0.0	0.0
BMCC	Big Moccasin Creek at Collinwood, Va.	0.43	0.97	0.03	0.0	0.0
BCMC	Buffalo Creek at Milligan College, Tenn.	0.62	0.75	0.25	0.0	0.0
BRHC	Bullrun Creek near Halls Crossroad, Tenn.	0.61	0.48	0.52	0.0	0.0
CCGC	Copper Creek near Gate City, Va.	0.39	0.89	0.11	0.0	0.0
LCCH	Lick Creek near Chapel Hill, Va.	0.96	0.39	0.61	0.0	0.0
LR-W	Little River at Wardell, Va.	0.21	0.97	0.03	0.0	0.0
LCWP	Long Creek near White Pine, Tenn.	0.30	0.59	0.41	0.0	0.0
MFHG	N. Fork Holston River at Groseclose, Va.	0.31	0.54	0.46	0.0	0.0
RC-O	Reedy Creek at Orebank, Tenn.	0.45	0.59	0.41	0.0	0.0
SC-D	Sewee Creek near Decatur, Tenn.	0.44	0.63	0.37	0.0	0.0
SC-B	South Chestuee Creek near Benton, Tenn.	0.45	0.27	0.73	0.0	0.0
SCFB	Stony Creek at Fort Blackmore, Va.	0.60	0.20	0.80	0.0	0.0

TABLE 16-12 (continued)

Code	Watershed	Forest	Independent Variables			
			C	S	I	U
	Blue Ridge:					
ACBC	Alarka Creek near Bryson City, N.C.	0.92	0.0	1.00	0.0	0.0
AC-H	Allen Creek near Hazelwood, N.C.	0.90	0.0	0.0	1.00	0.0
BC-S	Beetree Creek near Swannanoa, N.C.	1.00	0.0	1.00	0.0	0.0
CC-F	Cane Creek near Fletcher, N.C.	0.76	0.0	0.40	0.60	0.0
CC-B	Catheys Creek near Brevard, N.C.	1.00	0.0	0.48	0.52	0.0
CC-W	Connelly Creek at Whittier, N.C.	0.87	0.0	1.00	0.0	0.0
CR-C	Cullasaja River at Cullasaja, N.C.	0.89	0.0	0.0	1.00	0.0
DCBC	Deep Creek near Bryson City, N.C.	1.00	0.0	0.02	0.98	0.0
DREI	Doe River at Elizabethton, Tenn.	0.72	0.09	0.10	0.81	0.0
DRE4	Doe River near Elizabethton, Tenn.	0.75	0.04	0.11	0.85	0.0
H103	Hominy Creek at Candler, N.C.	0.71	0.0	0.05	0.95	0.0
MRMR	Mills River near Mills River, N.C.	0.97	0.0	0.86	0.14	0.0
NFCC	N. Fork Citico Creek near Tellico Plains, Tenn.	1.00	0.0	1.00	0.0	0.0
NFSR	N. Fork Swannanoa River near Grovestone, N.C.	0.91	0.0	1.00	0.0	0.0
NR-N	Nantahala River at Nantahala, N.C.	0.95	0.01	0.50	0.49	0.0
NCBC	Noland Creek near Bryson City, N.C.	1.00	0.0	0.91	0.09	0.0
NICE	North Indian Creek near Erwin, Tenn.	0.87	0.24	0.61	0.15	0.0
OR-B	Oconaluftee River at Birdtown, N.C.	0.96	0.0	0.27	0.73	0.0
OR-C	Oconaluftee River at Cherokee, N.C.	0.98	0.0	0.15	0.85	0.0
TC-T	Turtletown Creek at Turtletown, Tenn.	0.88	0.0	1.00	0.0	0.0
VR-A	Valley River at Andrews, N.C.	0.92	0.13	0.15	0.72	0.0
WFPW	W. Fork Pigeon River near Waynesville, N.C.	0.99	0.0	0.0	1.00	0.0
WRSG	Watauga River near Sugar Grove, N.C.	0.30	0.0	0.0	1.00	0.0

Test watersheds:

Code	Watershed					
	Mississippi Embayment:					
BC-H	Birdsong Creek near Holliday, Tenn.	0.72	0.0	0.0	0.0	0.0	1.00
CR-M	Clarks River at Murray, Ky.	0.21	0.0	0.0	0.0	0.0	1.00
	Highland Rim and Nashville Basin:					
BRCL	Big Rock Creek at Lewisburg, Tenn.	0.19	0.95	0.05	0.0	0.0
BC-W	Blue Creek near Waverly, Tenn.	0.60	0.73	0.27	0.0	0.0
WFMC	W. Fork Mulberry Creek near Booneville, Tenn.	0.46	0.93	0.07	0.0	0.0
WC-W	Wartrace Creek at Wartrace, Tenn.	0.24	0.84	0.16	0.0	0.0
	Cumberland Plateau:					
CF-W	Crooked Fork near Wartburg, Tenn.	0.80	0.0	1.00	0.0	0.0
SCST	Soddy Creek at Soddy, Tenn.	0.84	0.0	1.00	0.0	0.0
SBSM	Stanko Branch near Higdon, Ala.	0.84	0.0	1.00	0.0	0.0
	Valley and Ridge:					
BC-J	Brush Creek near Johnson City, Tenn.	0.23	0.88	0.12	0.0	0.0
CCHV	Cove Creek near Hilton, Va.	0.57	0.73	0.27	0.0	0.0
LC-M	Lick Creek at Mohawk, Tenn.	0.43	0.32	0.68	0.0	0.0
WCSC	White Hollow near Sharps Chapel, Tenn.	1.00	1.00	0.0	0.0	0.0
	Blue Ridge:					
JCCC	Jonathans Creek near Cove Creek, N.C.	0.67	0.0	0.26	0.74	0.0
LS-R	Little Santeetlah Creek near Robbinsville, N.C.	1.00	0.0	0.0	1.00	0.0
PR-H	Pigeon River near Hepco, N.C.	0.75	0.0	0.16	0.84	0.0

[a] After Betson and McMaster [14].

The model was calibrated on 66 watersheds in the Tennessee Valley as listed in Table 16-12. The drainage areas range in size from 5.5 to 764 sq. miles. The coefficients in the rating curve were determined by fitting the equation to concentration-discharge data from the 66 watersheds. The rating curves were established for 15 standard mineral constituents for each watershed. The average percent standard error in fitting was about 40%.

TABLE 16-13 Tennessee Valley Authority mineral water quality model test watershed simulations[a]

Watershed	Date	Q (cfs/ sq m)	Value Status	SiO₁	Fe	Ca	Mg
Little Santeetlah Creek near Robbinsville, N. C. (5.78 sq m) (BR)*	5/12/65	2.25	Observed	4.6	0.03	0.9	0.3
			Simulated	8.0	0.02	0	0.2
	8/ 3/64	0.865	Observed	5.2	—	1.2	0.2
			Simulated	9.1	0	0	0.2
Jonathans Creek near Cove Creek, N. C. (65.3 sq m) (BR)	5/ 2/57	3.23	Observed	11	0.01	2.0	1.2
			Simulated	7.8	0.01	3.2	1.0
	12/30/70	1.50	Observed	8.6	0.04	3.3	0.5
			Simulated	8.2	0.01	3.7	1.2
Cove Creek near Hinton, Va. (17.6 sq m) (VR)	4/ 4/68	2.73	Observed	3.5	0.04	44	6.1
			Simulated	5.0	0.04	34	6.2
	10/ 5/67	0.125	Observed	3.9	0.02	52	13
			Simulated	5.2	0.02	47	12
White Creek near Sharps Chapel, Tenn. (2.68 sq m) (VR)	3/29/72	4.10	Observed	5.8	0.04	28	13
			Simulated	3.7	0.06	42	5.7
	12/29/71	0.735	Observed	6.7	0.06	39	14
			Simulated	4.3	0.05	46	9.4
Stanko Branch near Higdon, Ala. (0.23 sq m) (CP)	12/30/69	43.9	Observed	2.9	0.09	2.3	0.7
			Simulated	5.4	0.02	3.6	0.5
	7/31/69	0.04	Observed	1.6	0.12	4.1	1.4
			Simulated	6.0	0.02	10	2.0
Soddy Creek at Soddy, Tenn. (49.0 sq m) (CP)	1/ 3/68	3.00	Observed	3.2	0.00	2.1	0.6
			Simulated	5.6	0.02	5.3	0.8
	7/ 2/68	0.020	Observed	4.1	0.06	10	1.7
			Simulated	6.1	0.02	11	2.3
W. Fork Mulberry Creek near Boonsville, Tenn. (17.4 sq m) (HR)	11/ 1/67	8.1	Observed	6.0	0.09	60	4.5
			Simulated	4.9	0.05	39	6.2
	9/ 4/68	0.046	Observed	4.5	0.07	48	6.5
			Simulated	5.1	0.02	63	18
Blue Creek near Waverly, Tenn. (24.8 sq m) (HR)	5/18/66	6.13	Observed	6.1	0.02	20	1.7
			Simulated	4.9	0.04	32	5.1
	12/ 7/65	0.375	Observed	5.4	0.01	41	4.3
			Simulated	5.1	0.03	42	9.3
Wartrace Creek at Wartrace, Tenn. (36.4 sq m) (NB)	12/12/66	3.02	Observed	5.9	—	64	5.7
			Simulated	5.5	—	40	7.8
	8/ 8/66	0.041	Observed	5.6	0.02	44	4.1
			Simulated	4.9	0.01	65	16
Big Rock Creek at Lewisburg, Tenn. (24.9 sq m) (NB)	12/12/66	3.41	Observed	5.5	0.01	56	5.9
			Simulated	5.4	0.03	44	8.5
	9/29/66	0.032	Observed	4.6	0.02	69	22
			Simulated	4.8	0.01	73	19
Birdsong Creek near Holliday, Tenn. (44.9 sq m) (ME)	12/ 9/66	3.85	Observed	6.4	0.02	26	1.7
			Simulated	6.8	0.60	2	0.7
	7/21/66	0.045	Observed	6.5	0.03	2.5	0.5
			Simulated	11.9	0.18	2.0	0.1
Clarks River at Murray, Ky. (89.7 sq m) (ME)	2/ 7/69	2.71	Observed	6.4	0.52	10	1.5
			Simulated	8.2	0.57	7	3.0
	8/15/67	0.11	Observed	9.6	0.40	8.5	0.2
			Simulated	9.9	0.18	6.5	0.6
Average absolute error × 100/ Average concentration				31	87	30	54

The model represents a first generation approach to simulating mineral loads since variations of the coefficients a and b would also be influenced by soil characteristics, erosion rates, the relations between ground and surface water, the ion content in rainfall, biological activity, and chemical buffering. These influences are very complex and are therefore not easy to model.

Constituent Concentration (mg/l except as noted)

Na	K	HCO₃	SO₄	Cl	NO₃	TDS	CaCO₃	Specific Conductance (Mm)	pH	Color (Pt-Co units)
0.6	1.2	5	2.2	0.5	0.1	7	3	13	6.4	5
1.0	0.5	3	0.5	0.7	0.2	15	1	0	6.7	7
0.7	0.5	6	0.4	1.0	0	11	4	12	6.5	5
1.2	0.5	4	0.5	0.7	0.2	17	1	0	6.8	8
1.4	0.9	10	0.5	1.3	1.8	25	12	27	6.2	5
1.6	0.9	10	4.3	1.4	0.9	28	12	48	6.8	8
2.6	0.7	8	2.7	3.5	—	32	10	38	6.2	5
1.7	0.9	12	4.2	1.4	—	31	14	52	6.9	8
2.0	1.1	139	12	3.5	1.3	116	134	241	7.9	5
1.3	1.4	125	5.5	1.9	2.3	126	111	205	7.7	3
3.2	—	200	10	4.5	1.7	198	174	338	8.0	3
1.7	—	187	6.5	2.4	2.0	171	170	280	7.9	3
0.5	0.8	137	3.7	1.5	0.07	135	128	202	7.8	8
0.3	0.9	166	1.3	1.1	1.3	140	127	186	7.7	1
0.7	0.5	178	4.5	1.0	0.6	145	154	250	8.0	3
0.5	1.6	180	1.8	1.4	2.7	160	155	218	7.8	1
0.7	1.4	2	5.6	2.5	0.3	28	8	34	5.0	30
0.8	0.5	7	4.3	1.1	0.2	22	11	33	6.7	8
1.3	2.7	13	5.4	5.0	0.0	33	16	52	6.1	5
2.1	1.9	31	8.6	1.7	0.0	55	32	93	7.0	7
0.7	0.0	2	5.6	1.5	0.2	13	8	26	6.2	5
1.1	0.8	13	5.6	1.3	0.1	31	17	50	6.8	8
1.8	1.1	9	28	1.0	0.0	44	32	94	6.3	5
2.4	2.2	36	9	1.7	0.0	60	36	103	7.0	7
1.4	2.1	179	13	2.0	2.7	205	168	320	7.8	5
1.3	1.2	140	5.6	2.0	2.7	140	119	231	7.9	2
2.0	3.1	183	3.2	4.0	4.2	172	148	280	7.4	5
1.8	4.3	254	6.6	2.8	3.0	223	241	365	8.3	2
1.5	2.1	65	4.6	1.5	0.9	65	56	120	7.1	5
1.1	1.1	113	5.0	1.8	2.2	115	98	185	7.6	3
1.8	1.7	145	5.0	2.5	1.7	147	121	250	7.1	5
1.4	2.2	161	6.0	2.2	2.1	152	145	246	7.8	3
2.0	1.7	204	12	4.0	5.1	204	183	332	8.2	5
1.9	1.7	142	8.3	2.7	3.5	149	130	266	8.0	4
6.5	7.4	142	10	7.5	9.3	167	128	295	7.5	8
1.8	4.2	273	7.0	2.9	1.5	227	244	392	8.3	2
2.3	2.7	176	13	6.0	—	188	164	310	7.5	5
2.0	1.7	157	8.4	2.8	—	162	142	289	8.1	3
4.2	6.7	240	16	7.5	—	253	208	392	7.9	5
1.8	4.8	306	6.7	3.0	—	251	280	427	8.5	2
2.3	4.2	74	13	3.5	1.1	97	73	162	7.7	20
1.8	0.7	5	7.0	2.3	0.02	30	9	25	6.4	12
1.5	1.3	10	2.0	2.0	1.0	30	8	29	6.6	3
5.1	2.3	6	1.3	2.9	0.05	29	5	21	6.6	5
2.8	2.7	20	16	4.5	3.3	57	31	87	6.5	35
3.1	1.3	15	14	3.5	0.7	50	28	102	6.7	15
4.2	3.1	24	4.4	6.0	2.5	47	22	78	6.7	7
4.0	2.4	22	2.5	3.2	0.4	49	18	88	6.8	5
40	45	30	46	44	66	22	31	23	7	55

[a] After Betson and McMaster [14].
[b] Drainage area and physiographic province.

The model was test applied to the 12 watersheds listed in Table 16-12. Simulations were made for actual samples taken as early as 1957. For each of the 12 test watersheds, the samples associated with the highest and lowest discharge were selected. The results of the tests are shown in Table 16-13. Generally, the simulators are close considering that the model is being used to reproduce mineral quality conditions at a point on a stream on a particular day 5–10 yr ago.

There are some large errors in simulating sulfate, perhaps because a portion of the sulfate load in the stream is brought in by rainfall. Simulations for Birdsong Creek were considerably in error for most constituents probably because the unconsolidated rocks in this province are widely heterogeneous, yet they are grouped into a single geologic variable in the model. The average absolute errors are shown at the bottom of each column and are comparable with the results obtained from the original model calibrations. Hence, if this model were to be used in the Tennessee Valley, average prediction errors of about $\pm 30\%$ should result for silica, calcium, bicarbonate, total dissolved solids, hardness, and specific conductance and $\pm 10\%$ errors should result for pH. The errors should be about $\pm 45\%$ for sodium, potassium, sulfate and chloride and about $\pm 55\%$ for magnesium. The errors for iron and nitrate should be about $\pm 100\%$. For many constituents, the maximum concentrations observed during short-term sampling over a range of streamflow conditions are usually two to five times the minimum. Simulations with the model at unsampled locations may be expected within this range.

Effects of Strip Mining on Water Quality—TVA and Stanford Models

EXAMPLE 16.4   A portion of the Crooked Fork watershed near Wartburg, Tennessee (50.3 sq miles) has been strip mined for coal. No samples had been taken before mining but samples were available during mining. The TVA daily flow model was used to simulate flows before mining began for two stormy days. The comparison between observed and simulated water quality constituents is shown in Table 16-14. This comparison indicates that strip mining has markedly increased the concentration of several constituents, particularly at low flows.

In the work of Herricks et al. [15] with the Stanford model (reported in Chapter 11), two aspects of water quality were investigated: (a) SO_4 concentrations and (b) sediment movement. The data generated by the Stanford model in Chapter 11 were used for both studies. Daily sequences were used for the sulfate model, while hourly discharge data for selected storm periods were necessary to predict incipient sediment movement. The period was between October 1970 and Semptember 1972. This period was of particular

TABLE 16-14 Example use of TVA daily flow model in evaluating the effects of strip mining on water quality in Crooked Fork near Wartburg, Tennessee[a]

Constituent concentration (mg/l except as noted)	July 12, 1967 (960 cfs)		July 29, 1968 (3 cfs)	
	Observed (after)	Simulated (before)	Observed (after)	Simulated (before)
S_1O_2	6.3	5.6	7.0	5.9
Fe	0.01	0.02	0.09	0.02
Ca	7.7	4.2	45	10
Mg	5.2	0.7	26	2
Na	2.0	1.0	20	2
K	1.2	0.6	3.7	1.8
HCO_3	10	9	2	30
SO_4	23	5	249	8
Cl	2.0	1.3	8.0	1.7
NO_3	0.0	0.3	0.5	0.0
TDS	48	25	409	54
$CaCO_3$	28	14	222	32
Sp. cond. (Mm)	81	42	530	96
pH	6.2	6.7	4.5	7.0

[a] After Betson and McMaster [14].

importance because detailed water quality and biological samples were obtained during this period and two major floods occurred that caused a severe impact on both stream physical and biological conditions. To better define the nature of this impact, detailed hydrographs were generated for each event from which the depth, velocity, and scour parameters could be determined for the peak flow.

Because sulfate is a conservative compound and is a good indicator of the occurrence and intensity of acid mine drainage, a model was structured to predict sulfate concentration in a given watershed based on background concentrations, known acid loadings, and stream discharge. The model was similar to a steady state conservative case model described by Thomann [16]. The model combined discharge and sulfate concentration from a tributary with the discharge and sulfate concentration of an upstream confluence, and gave the sulfate concentration for the discharge immediately downstream from the confluence. Because unpolluted tributaries have some sulfate in solution, they were assigned background sulfate concentrations of 20 ppm or less. Also, since mine drainage loading in acid tributaries is variable, these tributaries were assigned variable sulfate loads based on observed concentrations as they were found to be related to stream discharge. The simulated discharge values were used for all subbasins except Poplar Run where recorded data were used. Sulfate concentrations were obtained and adjusted according to values of water quality analyses made during their study. Sulfate loads for Champion Run and Poplar Run were reported to be 18,000 lb/day and 2000 lb/day, respectively.

Field observations showed that an extensive tailing or gob pile along the north bank of Champion Run produced significant amounts of acid mine drainage which resulted in high sulfate loading when rainfall caused runoff from this area. During dry periods or periods of light rainfall, the pyritic materials in the gob pile reacted with available moisture to produce large quantities of acid and sulfate compounds. This material was then flushed from the gob pile by high intensity rainfall events. Drainage during the early stages of the event were characterized by highly concentrated acid mine drainage, which was then followed by reduced acid drainage until the cessation of rainfall. To reproduce the effect of these concentrated loadings, they increased sulfate loads when rainfall exceeded 0.25 in.

EXAMPLE 16.5 Following this assumption, sulfate concentrations were predicted and compared with recorded data. During periods of moderate to high discharge, simulated sulfate concentrations compared favorably with values recorded during the recovery study. However, concentrations predicted during periods of low discharge were much higher than recorded values. This indicated that sulfate loadings decreased with groundwater flow. In an attempt to better model concentrations during the baseflow periods, modifications were made to allow sulfate loadings to decrease in proportion to streamflow. These results compared more favorably with recorded sulfate values.

Predicted sulfate loadings were, in general, very good. They are compared with the observed in Table 16-15. Because sampling occurred over a 1-to-5 day period, model values were selected for the midpoint of the week of sampling, and a second set was selected on the basis of closeness of fit of five values, with respective recorded values. Midpoint correlation values

TABLE 16-15   Correlation between recorded and predicted sulfate concentrations, Indian Creek Basin

Station No.	Midpoint (R)	Five-day best fit
3	0.69	0.94
4	0.74	0.89
7	0.50	0.84
10	0.35	0.69

varied between 0.35 and 0.75 while best fit values varied between 0.69 and 0.94. Considering the possible sources of error in the hydrologic model, and the loading values, the results were considered to be well within reason (see Table 16-15).

The sulfate values generated in this manner were recorded in relation to discharge in Fig. 16-14. The figure shows the relationship between acid drainage and stream discharge very clearly. The Champion Run tributary which was severely affected by acid drainage, showed very high sulfate concentration at low discharge. Although the dilution capacity of high discharges in this tributary was limited, low concentrations were often encountered because higher stream discharges not only diluted acid mine drainage but also prevented additional acid formation by rapid exchange of waters in contact with pyritic materials. The results were lower sulfate concentrations entering Indian Creek. These changes were not as radical at locations in the midstream region.

EXAMPLE 16.6   The Indian Creek drainage area studied by Herricks *et al.* [15] has been subjected to considerable mining, and as a consequence much of the drainage system has been degraded. To gain some insight on possible effects of high flows on the biota of the Indian Creek network they developed a routine utilizing sediment transport theory to predict the maximum size sediment particles that could be moved at a given flow stage. Only incipient motion was predicted and not the transport of the material scoured.

Two major floods which occurred on the watershed during the nine-month period between Semptember 1971 and June 1972 were selected for

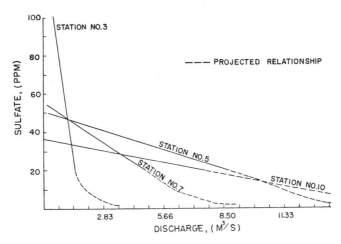

FIG. 16-14. Sulfate concentration as related to discharge for selected stations on Indian Creek. (After Herricks *et al.* [15].)

the study. The flood hydrographs were generated by the Stanford Watershed model.

The depths and velocities of flow are necessary to compute sediment movement. Water surface profiles were computed to provide flow characteristics for a given discharge through a known cross-section input data for sediment movement computations. Water surface profiles are computed by a solution of Eqs. (3-6) and (3-17) without the transient terms.

The output from water surface profiles included hydraulic radius R, and the slope of the energy gradient S_f from which the shear stress or tractive force was calculated:

$$\tau = \delta_w R S_f \qquad (16\text{-}15)$$

where δ_w is the specific weight of water. Using Shield's diagram to establish the relationship between Reynolds number and shear stress, the diameter of particles which may be moved by any tractive force are defined as

$$d = \tau/0.06(\delta_s - \delta_w) \qquad (16\text{-}16)$$

The largest diameter particle to be moved by the given stormflow was determined from Eq. (16-16). Even though the stream bottom of Indian Creek consisted mainly of sandstone and shale, estimates of the specific gravity of bottom materials were made to encompass a wide range of material types ranging from a mass density δ_s of 2.2 for salt, and a midrange value of 2.4, and 2.6, for nonporous sandstone. The data are summarized in Table 16-16.

The particle size movement was plotted against depth of flow and discharge in Figs. (16-15) and (16-16). From channel morphometry, estimates

TABLE 16-16 Maximum particle size moved at peak flow during the September 1971 flood at selected stations located within the Indian Creek drainage basin[a]

Station No.	Depth (ft)	Q (cfs)	Diameter moved with given a_s (in.)		
			2.4	2.4	2.6
Station No.	(ft)	(cfs)	2.2	2.4	2.6
2a	12.86	9009	2.00	1.80	1.50
4	15.52	12584	1.10	0.96	0.84
5	14.64	13701	2.00	1.80	1.50
7	14.07	16431	3.90	3.20	2.90
8	22.44	20115	0.80	0.70	0.60
10	14.72	23692	7.10	6.00	5.30
10a	13.20	26139	1.30	1.10	0.90

[a] After Herricks et al. [15].

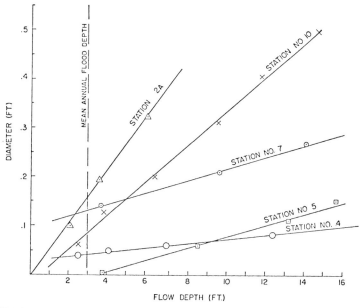

FIG. 16-15. Particle size movement as related to flow depth, Indian Creek. (After Herricks *et al.* [14].)

were made of bank full discharge and then related to the mean annual flood. When these results were compared with mean values from the hydrologic modeling, their results were favorable.

The mean annual flood depth was found to be 3 ft. At this depth the mainstream stations showed variable movement. The most important fact relates to the movement potential range below the mean annual flood. In each instance the largest size moved was estimated to be less than 1.8 in. From the standpoint of biological impact, the mean annual flood has the potential to move all but the largest stream fauna. They concluded that higher streamflows can alter the bottom and cause severe disruption of biological function [15]. A second relationship is shown in Fig. 16-16 which relates particle size to discharge. Station 2a was a narrow station with moderate gradient; with small changes in discharge, a large variation in particle size movement occurred. The other mainstream stations had a wide channel morphometry, and the variability of size of materials moved was not as pronounced with discharge. Downstream stations did show the greatest size of movement. This may be attributed to higher discharge volumes with each station downstream receiving proportionally higher discharge loads. Again, size movement stayed within 0.2 ft, demonstrating the potential impact of higher streamflows.

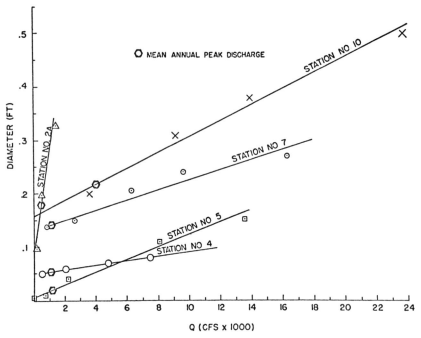

FIG. 16-16. Particle size movement as related to discharge, Indian Creek. (After Herricks *et al.* [14].)

Problems

16-1. Why model stormwater quality?

16-2. Discuss the different seasonal effects on total accumulation on streets.

16-3. Outline an approach to a two-year study to assess the impact of urban development on stormwater quality. Include the data that would be required, and discuss how mathematical modeling can be useful.

16-4. Discuss the representativeness of the data in Table 16-2–16-5, i.e., how representative will the data be 5 yr after the study, considering the constantly changing environmental and land use conditions.

16-5. Rework Example 16.2 when the average slope is 7% and the slope length is 560 ft.

16-6. How much will the sediment yield from a 10-acre construction site in Knoxville, Tennessee be increased/decreased if the grade of a 500-ft slope changed from 1.5 to 3%? The soil erodibility factor is 0.34 and the design storm is the 5-yr storm.

16-7. Derive Eq. (16-8).

16-8. Rework Example 16.3 for an industrial area with 45 miles of streets and a total solids loading of 900 lb/curb mile. Using the hydrographs in Fig. 16-10.

16-9. Rework Example 16.3 for an urban area with 30 miles of streets and a BOD loading of 10 lb/curb mile. Use the hydrographs in Fig. 16-10, and assume that total area runoff is twice the street runoff at each point on the hydrograph. Compare the instantaneous concentrations with local drinking water standards.

16-10. Discuss the ultimate effects of stormwater detention basins on the quality of receiving water bodies.

16-11. What is the usefulness of regression in stormwater quality modeling? When can the model developed for one catchment be applied to another catchment?

References

1. US Environmental Protection Agency, Water quality management for urban runoff, EPA 440/9-75-004, December 1974.

2. Sartor, J. D., and Boyd, G. B., Water pollution aspects of street surface contaminants, *Environ. Protection Technol. Series,* EPA-R2-72-081, November 1972.

3. Shaheen, D. G., Contributions of urban roadway usage to water pollution, *Environ. Protection Technol. Series,* EPA-600/2-75-004, April 1975.

4. Wischmeier, W. H., A rainfall erosion index for a universal soil-loss equation, *Soil Sci. Soc. Amer. Proc.* **23,** 246–249 (1959).

5. Musgrave, G. W., The quantitative evaluation of factors in water erosion, a first approximation, *J. Soil Water Conserv.* **2,** 133–138 (1947).

6. Ellison, W. D., Some effects of raindrops and surface flow on soil erosion and infiltration, *Trans. Am. Geophys. Union* **26,** 415 (December 1945).

7. Wischmeier, W. H., and Smith, D. D., Predicting rainfall-erosion losses from cropland East of the Rocky Mountains, US Dept. of Agr. Handbook No. 282, Agr. Res. Service, Washington, D.C., 1965.

8. Univ. of Cincinnati, Urban runoff characteristics, Interim Report to Environmental Protection Agency Water Quality Office, Oct. 1970.

9. Roesner, L. A., Quality aspects of urban runoff, "Short Course on Applications of Stormwater Management Models," Dept. of Civil Eng., Univ. of Massachusetts, August 1974.

10. Barkdoll, M. P., An analysis of urban stormwater quality and the effect of dustfall on urban stormwater quality; East Fork of Third Creek, Knoxville, Tennessee, M.S. Thesis, Dept. of Civil Eng., Univ. of Tennessee, Knoxville, Tennessee, 1975.

11. Colston, Newton, V., Jr., Characteristics and treatment of urban land runoff, *Environ. Protection Technol. Series,* EPA-670/2-74-096, (December 1974).

12. US Environmental Protection Agency, Stormwater management model. II. Verification and testing, *Water Pollution Control Res. Series,* 11024DOCO8/71, (1971).

13. Betson, R. P., Upper Bear Creek experimental project—a continuous daily streamflow model, Tennessee Valley Authority, Knoxville, Tennessee, February 1972.

14. Betson, R. P., and McMaster, W. M., Nonpoint source mineral water quality model, *J. Water Pollution Control Fed.* **47,** No. 10, 2461–2473 (1975).
15. Herricks, I. E., Shanholtz, V. O., and Contractor, D. N., Hydrologic and water quality modeling of surface water discharges from mining operations, Virginia Poly. Inst. State Univ., Blacksburg, Virginia, January 1975.
16. Thomann, R. V., "Systems Analysis and Water Quality Management," Environ. Sci. Services, New York, 1972.

Chapter 17 | Development of Stormwater Quality Indices

17-1 Introduction

The purpose of a water quality index is to facilitate the description of the quality of water with a number or a simple group of numbers. This necessitates a blending of the values of the various indicators such as pH, and dissolved oxygen in a manner which will define the state of the quality of the water. However, it is very important to understand that no such index is complete without being predicated upon a value judgment pertaining to the intended use of the water.

The development of water quality indices is also desirable from the point of view that the total number of constituents to be measured is placed at a minimum. Sampling and water analysis is very expensive and the data generated can be overpowering from the standpoint of the analyst. Further, in order to achieve the goal of relating stormwater quality to the associated hydrologic and land use factors, it is first necessary to develop a water quality index in the form of a vector(s). This provides a compact form of the water quality and permits an efficient correlation with the causative factors.

There has been little work done in the development of water quality indices in general, and almost nothing reported in the development of stormwater quality indices. Two basic approaches are presented herein. In the first approach, a water quality index was developed on the basis of a value judgment that the use of the water would be primarily for recreation. In the second approach, principal components analysis was utilized to filter out statistically redundant water quality variates, to identify the nature of each component and to correlate the components with stormwater discharge.

17-2   "Consumers' Water Quality Index"

The most important problem involved in developing a water quality index is that there are many uses for water and the associated quality of water demanded for each purpose varies widely. The desirability of the value associated with a water quality variable often depends upon the use to which the water is put.

Walski and Parker [1] developed a water quality index primarily for use in establishing the desirability of water for recreational use. Eleven quality variables were combined as a geometric mean to form their water quality index. The geometric mean was selected over the arithmetic sum because it is less affected by extremes in water quality values. A sensitivity function for each quality constituent was generated from information reported in the literature. The values of the sensitivity functions which were considered to be perfect, good, poor, and intolerable were selected and each of these values were assigned the number 1.0, 0.9, 0.1, and 0.01, respectively. At this point, linkage of values of each of the quality constituents with the assigned values of the sensitivity functions provided a basis for generating values of the water quality index. The developed index was test applied to actual water bodies and it was shown that the index did reflect the quality of the water bodies.

One of the main deficiencies associated with use of the index was its failure to agree with observable water quality in situations where allegedly interrelated variables such as suspended solids, turbidity, and color increased simultaneously. Hence, regardless of the use that an index is put to or of the particular mathematical form that the index has, interrelations among the data may very well cause the index to produce illogical and erroneous predictions. These problems may be offset by application of multivariate statistics. First, statistically redundant variates may be filtered out; and second, the index can be made up of components which are not statistically interrelated, rather than the original water quality variates. It therefore follows that there are two types of indices to be developed rather than only one.

The type index which Walski and Parker [1] have developed is very much needed and their work has provided a fundamental framework for development of water quality indices for application to a variety of water uses. However, prior to formulating the use index and the associated sensitivity functions, it seems logical to first develop indices of physical, chemical, and biological variables in the form of components.

17-3 A Multivariate Approach

Mahloch [2] presented a detailed analysis of water quality of the Pascagoula River in Mississippi. The data included the 17 variables listed in Table 17-1 which were measured at 36 streamflow stations along the river. Principal components analysis, as described in Chapter 9, was applied to the data with the water quality variables combined in linear form.

The results of the Varimax rotation are shown in Table 17-2. Shown in the table are the first six components forced out which represent 76% of the total variance of the original data. The first component is related to the dissolved solids concentration and composition of the river. The second component appears to represent the degree-of-pollution at any sampling point. Since this component is highly correlated with BOD and fecal coliform, it may be inferred that the major pollution source for the river basin

TABLE 17-1 Water quality variables in Mississippi study[a]

Variable	Variable
1	Specific conductance ($\mu\Omega$)
2	Temperature (°C)
3	Flow (cfs)
4	Dissolved oxygen (mg/l)
5	pH
6	Color (P–C units)
7	Chloride (mg/l)
8	Dissolved solids (mg/l)
9	Ammonia-N (mg/l)
10	Nitrite-N (mg/l)
11	Organic-N (mg/l)
12	Nitrate-N (mg/l)
13	Total phosphate (mg/l)
14	Sodium (mg/l)
15	BOD (mg/l)
16	Total coliform (log/100 ml)
17	Fecal coliform (log/100 ml)

[a] After Walski and Parker [1].

TABLE 17-2   Varimax rotation of principal components solution of Pascagoula River data

	Component					
Variable	1	2	3	4	5	6
Specific conductance	*0.795*	0.243	−0.071	0.050	0.000	0.147
Chloride	*0.891*	−0.141	0.084	−0.008	−0.056	−0.009
Dissolved solids	*0.919*	−0.027	−0.010	−0.050	−0.228	−0.196
Sodium	*0.931*	−0.017	0.117	−0.012	−0.204	−0.188
BOD	0.156	*0.822*	0.001	0.113	−0.147	0.158
Total coliform	−0.084	*0.754*	0.484	−0.096	0.010	−0.199
Fecal coliform	−0.167	*0.864*	0.099	−0.072	−0.040	−0.250
Ammonia-N	0.031	0.163	*−0.671*	−0.185	0.242	0.032
Nitrite-N	0.061	0.291	*0.712*	−0.180	0.101	0.002
Temperature	−0.030	0.014	0.361	*0.727*	0.254	−0.232
Flow	−0.012	−0.045	−0.089	*0.862*	−0.120	0.084
Color	0.364	−0.161	0.117	−0.112	*−0.686*	−0.282
Organic-N	0.207	0.328	−0.019	0.152	*−0.796*	0.207
Total phosphate	0.031	0.602	0.006	−0.184	*−0.622*	−0.173
Dissolved oxygen	−0.302	0.467	0.155	−0.107	0.006	*0.676*
Nitrate-N	0.021	0.030	0.196	0.006	−0.058	−0.807
pH	0.169	0.282	0.487	0.247	0.072	−0.216

is domestic in origin. The third component represents two nitrogen sources for the river. These variables have opposite correlations (directional cosines) with this component because of the dipolar positions of ammonia and nitrate in the nitrification process. The fourth component represents the hydraulic state of the river where the seasonal effect on flow appears as the positive correlation between temperature and flow. The fifth component represents general nutrient load of the river. The relation between nutrients and color can be accounted for by either sources of nutrients in the runoff or by the algae growth in the river. The sixth component represents the oxygen state in the river. Dissolved oxygen and nitrate are inversely related here because of nitrification.

At this point, a consumer's water quality index could be developed for the Pascagoula using the results of Mahloch's analysis. Since the components and the interrelations have been identified and the quantified, sensitivity functions for the components can be developed in the manner suggested by Walski and Parker [1]. Further, since conductance, chloride, dissolved solids, and sodium are highly interrelated, three of these original variates may be filtered out and the process is thus simplified.

Although this analysis is of a receiving water body and not of stormwater per se, it is one of the most complete multivariate analyses of water quality data.

17-4 Principal Components Regression

Barkdoll [3] reported an analysis of stormwater quality as it relates to storm runoff volume in a 1.6 sq. mile watershed in Knoxville, Tennessee. The Third Creek watershed is an older urban area with 14% of the land use in industrial activities. The major industry in the watershed is a scrap yard and foundry located adjacent to the main channel. Other industries are two trucking firms, a lumber yard, a porcelain plant, a linen service, a building materials supplier, and a wood veneer plant. Most of the industry is located adjacent to either the main channel or one of the minor channels. Approximately 71% of the area is residential. Most houses are older, single dwelling units. The general economic status of the area is low income. One housing project is located in the area consisting of multiple dwelling units. Approximately 1% of the land is used for commercial purposes. The remaining 14% is undeveloped. The undeveloped areas are primarily steep forested ridges. The majority of the undeveloped land is located on Sharp's Ridge which is in the headwaters. The length of the main channel is 8800 ft, and the tributaries add up to 6400 ft. The average slope of the main channel is 2.3% and the maximum elevation drop in the watershed is approximately 300 ft.

While Knoxville in general is located in a karst region, no sink holes are found in the watershed. The study area is primarily underlain by shale foundations, although some sections have sandstone and limestone.

There are a total of 22.4 miles of roads located in the study area. The population density of the area is approximately 12 persons/acre. Very little construction activity took place during the study period, hence, land use remained relatively constant. There is one major traffic artery within the area with a length of 7000 ft. The study area is located in the fork of two major interstate highways, so that three sides are in a relatively close proximity to these interstate highways and they are very heavily traveled. There are also 2.2 miles of railroads located within the watershed. These tracks are not extensively used.

Water quality data were collected by a proportional flow collection system such that only flows greater than 2 cfs were sampled. A maximum of 56 one-half liter samples were collected between servicings. Each batch of samples was composited and then analyzed. Values obtained represented average concentrations over the sampling period.

Dust fall data were collected in a single dust fall jar located approximately one mile from the streamflow and raingauge station. The dust fall jar used was 7.7 in. in diameter, positioned 4 ft off the ground, and equipped with a hard ring. The dust fall samples were analyzed for the same constituents as water quality. The relations between dust fall and water quality were reported in Chapters 15 and 16.

In selecting data to be used for this study, certain elimination guidelines were used. Seventy-three storms were sampled for water quality data during the period of October 1971 to October 1974. The hydrographs for each storm were studied and 47 storms were selected for preliminary investigations. Storms deleted were those in which (1) several storms occurred during the sampling period, (2) only part of the hydrograph was sampled over, (3) samples that were taken were not weighted with flow, and (4) samples were removed a considerable time after the storm occurred. Although some of the remaining storms were not single events, these storms were deemed representative of very wet periods entailing complex storms which occur in this area during the late winter and early spring. BOD was dropped from the data matrix because of the problems encountered by other researchers.

The data matrix consisted of 47 storm events analyzed for 20 water quality variables plus storm runoff volume. Principal components analysis revealed no significant interrelations amongst the water quality variates, so much so that the original variates nearly formed an orthogonal data set. The highest simple correlation coefficient was between chloride and sodium and was equal to 0.63. Further, a linear principal components regression model was used to correlate storm runoff volume as the dependent variable with a linear sum of the water quality variates. Only 25% of the variance of storm runoff volume was explained.

There appears to be little explanation for the lack of correlation amongst the water quality variates and with the storm runoff volumes. Our knowledge of the storm runoff and pollutant removal process does however, indicate that the rainfall excess intensity hyetograph significantly controls these processes. Hence, composite storm samples for small flashy watersheds may very well damp out statistically perceptible variations and interrelations in stormwater quality. For larger watersheds and especially in large receiving water bodies such as in the Pascagoula River study of Mahloch, the effects of rainfall intensity are damped out and variations in stormwater quality averages do not filter out variations and interrelations amongst the variates.

17-5   Conclusions

Much work needs to be done in the development of stormwater quality indices. Very few studies have been reported which attempt the development of indices. However, many studies are underway which will probably contribute to the state of the art. The reader is advised at this point to keep up with the open literature on this subject.

Problems

17-1. What is a water quality index and what is its utility?

17-2. How are value judgments involved in the development of water quality indices?

17-3. How may multivariate statistics be useful in developing a water quality index?

17-4. Develop your own water quality index for a water body or watershed in your locale and test apply it.

17-5. What are sensitivity functions as used in a water quality index?

17-6. Discuss the effects of rainfall intensity on any alleged interrelations amongst stormwater quality variates.

References

1. Walski, T. M., and Parker, F. L., Consumers water quality index, *J. Environ. Eng. Div.*, *Amer. Soc. Civil Engr.* **100,** No. EE3, 593–611 (1974).
2. Mahloch, J. L., Multivariate techniques for water quality analysis, *J. Environ. Eng. Div.*, *Amer. Soc. Civil Engr.* **100,** No. EE5, 1119–1132 (1974).
3. Barkdoll, M. P., An analysis of urban stormwater quality and the effects of dustfall on urban stormwater quality: East Fork of Third Creek, Knoxville, Tennessee. M.S. Thesis, Dept. of Civil Engr., Univ. of Tennessee, Knoxville, Tennessee, 1975.

APPENDIX 1

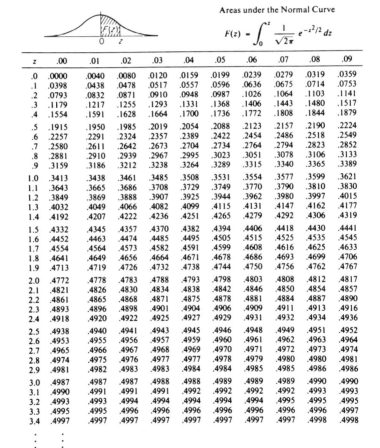

Areas under the Normal Curve

$$F(z) = \int_0^z \frac{1}{\sqrt{2\pi}} e^{-z^2/2} dz$$

z	.00	.01	.02	.03	.04	.05	.06	.07	.08	.09
.0	.0000	.0040	.0080	.0120	.0159	.0199	.0239	.0279	.0319	.0359
.1	.0398	.0438	.0478	.0517	.0557	.0596	.0636	.0675	.0714	.0753
.2	.0793	.0832	.0871	.0910	.0948	.0987	.1026	.1064	.1103	.1141
.3	.1179	.1217	.1255	.1293	.1331	.1368	.1406	.1443	.1480	.1517
.4	.1554	.1591	.1628	.1664	.1700	.1736	.1772	.1808	.1844	.1879
.5	.1915	.1950	.1985	.2019	.2054	.2088	.2123	.2157	.2190	.2224
.6	.2257	.2291	.2324	.2357	.2389	.2422	.2454	.2486	.2518	.2549
.7	.2580	.2611	.2642	.2673	.2704	.2734	.2764	.2794	.2823	.2852
.8	.2881	.2910	.2939	.2967	.2995	.3023	.3051	.3078	.3106	.3133
.9	.3159	.3186	.3212	.3238	.3264	.3289	.3315	.3340	.3365	.3389
1.0	.3413	.3438	.3461	.3485	.3508	.3531	.3554	.3577	.3599	.3621
1.1	.3643	.3665	.3686	.3708	.3729	.3749	.3770	.3790	.3810	.3830
1.2	.3849	.3869	.3888	.3907	.3925	.3944	.3962	.3980	.3997	.4015
1.3	.4032	.4049	.4066	.4082	.4099	.4115	.4131	.4147	.4162	.4177
1.4	.4192	.4207	.4222	.4236	.4251	.4265	.4279	.4292	.4306	.4319
1.5	.4332	.4345	.4357	.4370	.4382	.4394	.4406	.4418	.4430	.4441
1.6	.4452	.4463	.4474	.4485	.4495	.4505	.4515	.4525	.4535	.4545
1.7	.4554	.4564	.4573	.4582	.4591	.4599	.4608	.4616	.4625	.4633
1.8	.4641	.4649	.4656	.4664	.4671	.4678	.4686	.4693	.4699	.4706
1.9	.4713	.4719	.4726	.4732	.4738	.4744	.4750	.4756	.4762	.4767
2.0	.4772	.4778	.4783	.4788	.4793	.4798	.4803	.4808	.4812	.4817
2.1	.4821	.4826	.4830	.4834	.4838	.4842	.4846	.4850	.4854	.4857
2.2	.4861	.4865	.4868	.4871	.4875	.4878	.4881	.4884	.4887	.4890
2.3	.4893	.4896	.4898	.4901	.4904	.4906	.4909	.4911	.4913	.4916
2.4	.4918	.4920	.4922	.4925	.4927	.4929	.4931	.4932	.4934	.4936
2.5	.4938	.4940	.4941	.4943	.4945	.4946	.4948	.4949	.4951	.4952
2.6	.4953	.4955	.4956	.4957	.4959	.4960	.4961	.4962	.4963	.4964
2.7	.4965	.4966	.4967	.4968	.4969	.4970	.4971	.4972	.4973	.4974
2.8	.4974	.4975	.4976	.4977	.4977	.4978	.4979	.4980	.4980	.4981
2.9	.4981	.4982	.4983	.4983	.4984	.4984	.4985	.4985	.4986	.4986
3.0	.4987	.4987	.4987	.4988	.4988	.4989	.4989	.4989	.4990	.4990
3.1	.4990	.4991	.4991	.4991	.4992	.4992	.4992	.4992	.4993	.4993
3.2	.4993	.4993	.4994	.4994	.4994	.4994	.4994	.4995	.4995	.4995
3.3	.4995	.4995	.4996	.4996	.4996	.4996	.4996	.4996	.4996	.4997
3.4	.4997	.4997	.4997	.4997	.4997	.4997	.4997	.4997	.4998	.4998
⋮	⋮									
4.0	.499968									

Index

A

Accumulation of pollutants, 306–310
Acid mine drainage, 215, 230
Algae growth, 348
Available water capacity, 33

B

Baseflow, 3, 197, 261
"Black box," 7, 160, 174, 181, 182

C

Capacity rates, 31
Combined sewer overflows, 120–122, 325–329
Components, 174–185, 347, 348
Conservation of mass, 22, 43, 117
Conservation of momentum, 42–45
Convolution, 136, 140
Correlation
  coefficient, 111, 171, 174, 178, 180, 182, 186, 204, 224, 250
  coefficient of determination, 251
  coefficient of variation, 203
  matrix, 178, 182, 183
Covariance, 175, 179
Culverts, 114–119

D

Darcy's law, 23, 25
Depression storage, 19, 21
Diffusivity, 25
Directional cosine, 176–180, 188
Double triangle model, see TVA stormwater model

E

Effects of forest cutting on stormwater, 207–209
Effects of reclaiming denuded land on stormwater, 254
Effects of semi-karst topography on stormwater, 255
Effects of storm sewers on stormwater, 274, 324
Effects of strip mining on streamflow, 215, 236, 336–342
Effects of urbanization on stormwater, 122–124, 166, 209–212, 274–276, 324–328
Eigenvalue, 179, 184, 189
Eigenvector, see Components and Principal component analysis
EPA model, 9, 14, 120–122
Equilibrium storage, 88–93, 145
Equivalent length, 141–146

Equivalent uniform plane, 135, 140
Evaporation, 21, 229, 232, 242
Evapotranspiration, 19, 37, 219, 227, 239
"Eyeball" best fit, 11, 160

F

Flow frequency analysis
  best fit criteria, 266–271
  central limit theorem, 272
  central tendency, 270
  class size, 265
  confidence intervals, 272
  cumulative distribution, 265, 272
  design risk, 277
  hypothesis testing, 273
  method of moments, 265
  nonlinear least squares, 270
  outliers, 270, 272
  plotting positions, 265
  probability density function, 264
  skewness, 265
Froude number, 46, 79

G

Gravity, 23, 45
Green and Ampt model, 27
Groundwater, 218, 227

H

High speed digital computer, 161, 164
Holtan–Overton model, 33
Horton infiltration model, 31
Hydraulic conductivity, 23, 99
Hydraulic head, 23
Hydroscopic water, 33
Hysteresis, 26

I

Infiltration, 19, 21, 22, 99, 104, 227, 239
Infiltrometer, 32

Initial abstraction, 36
Instantaneous response function, 132–136,
  161–163, 194, 210
Interception, 19, 20, 21, 217
Interflow, 3, 227

K

Kinematic flow
  cascade of planes, 76, 89, 104, 134, 141,
    150–153
  Chezy formula, 59, 106
  converging surface, 75, 91
  depth profile, 68–70, 88
  equilibrium, 63
  falling hydrograph, 69, 116
  kinematic flow number, 59
  kinematic shock, 77–79
  lag modulus, 93–96, 109, 120, 211
  lagtime, 63, 88, 132–145, 163, 211
  overland flow, 58–79, 161
  plane flow, 62–70, 88, 145–150
  rising hydrograph, 61–68, 116
  storm sewer flow, 81, 119
  streamflow, 54, 79–81
  time to equilibrium, 68
  turbulent flow, 61, 62, 66, 147
  viscous flow, 61, 62, 66, 105
  V-shaped watershed, 70–74, 90
  wave speed, 87

L

Lagrangian multiplier, see eigenvalue
Loadograph, 303, 320, 321, 323–343

M

Manning's formula, 46, 47
Method of characteristics, 47–49, 69, 74–78,
  80, 87
Minshall's example, 130
Model complexity, 10, 193
Model constraints, 10, 138
Model linkage, 10, 210–212

Model objective, 4, 134, 169, 181, 193, 230
Model reliability, 4, 213, 227, 233, 243, 256, 264, 272
Model verification, 4, 8, 9, 11, 100–108, 146–151, 159, 204, 223, 228, 230, 232
Model process, 4, 160, 169, 192, 230, 238, 241, 321–323
Moving rainstorms, 152

N

Newton–Raphson method, 111
Normalized variates, 175–185
Numerical methods, 49–54

O

Objective best fit, 8, 101, 104, 134, 160, 162, 166, 169–189, 265
Objective function, 172–174, 187, 268
Operational bias, 12
Optimization techniques
    linear least squares, 169–172, 322
    nonlinear least squares, 172–174, 322
    pattern search, 187, 200, 220
    regression analysis, 261, 274, 321–323
    Rosenbrock's method, 186, 241
    stepwise multiple regression, 186, 246, 322
Orthogonal system, 175–185, 186

P

Parameters, 6, 8, 169, 172, 174, 180, 186, 193, 200, 203, 216, 228, 238, 239
Partial area contribution, 3, 194
Physical significance, 12, 172, 181
Policy space, 187
Pollutograph, 303, 320, 321, 323–343
Principal components analysis, 174, 178, 347–350
Probability
    conditional, 262, 277
    density function, 264
    simple, 262, 277
    statistical inference, 264

R

Rainfall excess, 19, 60, 109, 130, 194, 211, 350
Rational method, 87, 110
Removable pollutants
    EPA model, 319
    impervious areas, 318
    mineral loads, 328, 330–336
    sediment yield, 311–316
Return period, 109, 262–264
Reynolds number, 61, 105
Richard's equation, 25, 98
Routing, 117–119, 135, 137–140, 219
Routing function, 137–140

S

Saturated flow, 23
S-curve, 135, 144
Sensitivity analysis, 13, 238–245, 265
Site characteristics, 8, 182, 208, 227, 232, 246, 249
Soil moisture characteristic, 25–99
Soil moisture storage, 218, 227
Soil physics, 22
Standard deviation, 175, 273
Standard error of estimates, 111, 135, 200, 262
Stanford model, 13, 165, 227–233, 336–342
Stationary processes, 262
Statistical interrelations, 170, 174, 180, 182, 184
Stochasticism, 4
Storm hydrograph, 102–108, 125, 204, 208, 212, 225, 233, 252–255
Stormwater detention basin, 114–119
Stormwater models
    ARS, 165
    comparisons, 125, 167
    components, 9, 19
    deterministic, 7, 8, 10, 98–126, 159, 212
    DIHM, 152
    distributed, 10, 162, 166, 193
    EPA–SWMM, 9, 10, 14, 120–122, 318, 324
    linear and nonlinear, 6, 129, 161–164, 166, 174
    lumped, 7, 162, 193
    memory, 6,

MIT, 122–124, 126
Nash, 161
optimization, 8, 105, 169–189, 200, 220, 238, 241
Overton and Tsay, 108–119
parametric, 7–9, 11, 159–167, 212
Purdue, 166
purely random, 7, 159
regionalized, 8, 13, 159, 181, 193, 203, 246–250
simulations, 9, 100–126, 146–153, 166, 224, 250
Smith and Woolhiser, 98–108
Stanford, 13, 165, 227, 233
state, 6
stochastic, 7, 159, 261
time-invariant, 7, 194
TVA, 8, 13, 14, 166, 189, 192–212, 215, 328
University of Cincinnati, 125
US Geological Survey, 13, 165, 187, 238
VRM, 129–154
Stormwater pollutants
  bacteria, 289, 297, 300
  BOD, 297, 298, 300, 305, 308, 323, 327, 347, 350
  COD, 287, 293, 297, 300, 305, 308, 323
  dissolved oxygen, 289, 348
  dustfall, 305, 308, 319, 349
  erosion, 296
  fecal coliform, 299, 310, 347
  fertilizers, 296
  first-flush, effect, 287
  heavy metals, 288, 290, 299, 308, 321, 323
  herbicides, 296
  nutrients, 297, 290, 300, 308, 323, 347
  oil and petroleum products, 290
  organic matter, 288, 294, 323, 347
  pesticides, 289, 290, 294
  point and nonpoint sources, 287, 291, 298
  sediment, 287, 289, 294, 297, 298, 300, 308, 323, 327, 347
  sulfate, 336–342
Stormwater quality
  agriculture, 294, 298, 328–336
  forests and woodlands, 300

  mining, 298–300, 336–342
  urban, 291–294
Stormwater quality indices
  "Consumers' Water Quality Index," 346
  multivariate model, 347
  sensitivity function, 346
Superposition, 6, 136, 142
System, 5, 129
Systems analysis, 5

T

Tension, 23, 99
Time of concentration, 87–96, 109, 166, 211
Time series analysis, 261
Trade-off diagram, 11, 165
Transpiration, 39
TVA daily flow model, 215–227, 328–342
TVA models, 8, 13, 14, 193, 215
TVA stormwater model, 193–212

U

Unit hydrograph, 6, 130, 161
University of Cincinnati model, 34
Unsaturated flow, 23, 99
Urban planning, 9, 122, 192
US Geological Survey model, 13, 165, 187, 238

V

Variable, 6, 172, 174, 182, 186
Variable response model, 131–136
Variance, 171, 178, 185, 203, 247
VARIMAX rotation, 177, 347

W

Waves
  dynamic, 45–47, 80
  kinematic, 45–47, 80, 87

A 6
B 7
C 8
D 9
E 0
F 1
G 2
H 3
I 4
J 5